W9-BFO-008

YOUR HORSE

The Illustrated Handbook to Owning and Caring for Your Horse

By Carolyn Henderson, author of the *Horse & Pony Book*

Edited by Fran Lynghaug, author of

The Official Horse Breeds Standards Guide

Voyageur Press

CONTENTS

4 Health and first aid 88

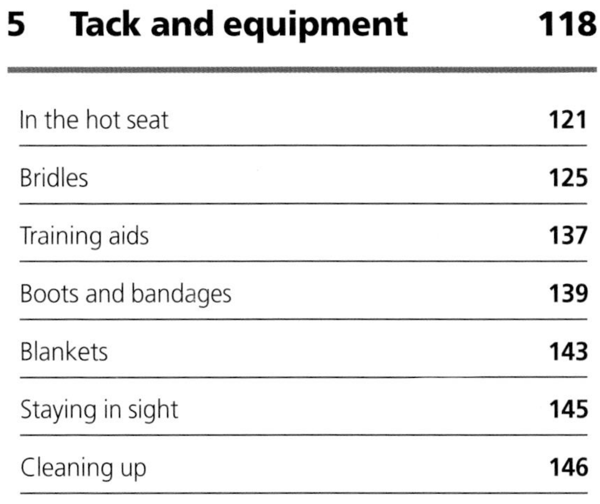

5 Tack and equipment 118

6 Buying a horse 148

7 Appendices 162

Introduction

People and horses have a unique relationship. Over thousands of years, we have used them to work the land, as a means of transport, in warfare, and in sport. Today, most horses are kept and ridden for competition and—just as important—for pleasure, but there are still parts of the world where families depend on their horse for survival.

Horses demand admiration, respect, and affection, from those who know them and those who admire them from afar. Many people, adults as well as children, find that learning to ride and look after a horse is rewarding and fun and some achieve their ultimate ambition by becoming horse or pony owners.

This book helps you understand the what, why, and how of horses—what they are, why they are so special, and how to understand and care for them. It's called *Your Horse: The Illustrated Handbook to Owning and Caring for Your Horse* because it will guide you through your relationship with horses. Perhaps you're reading it because you or someone else in your family is learning to ride or you're thinking of buying a horse or helping to look after one. Alternatively, you might simply want to find out why horses are so fascinating!

However, it isn't a users' or owners' manual such as you'll find with, say, a washing machine or a car. Horses have wonderfully generous temperaments and years of domestication have made them willing companions. They still have minds of their own, though, which is what makes building relationships with them so rewarding.

The more you learn about horses, the more you realize that you never stop learning. But by the time you get to the end of this book, you'll have a basis of knowledge on which to build. Just remember—even though owning a horse isn't always possible, everyone can enjoy and appreciate them.

You'll find that where an individual horse is referred to, it is as "he" rather than "she" unless the content dictates otherwise. This isn't being sexist, but is simply to avoid convoluted sentences!

PART 1

WHAT IS A HORSE?

Talking horse

The horse world has a language all of its own, and throughout this book you'll find words and phrases that are explained in context. It will help to understand a few of them right from the beginning.

Horses and ponies are part of the same group, Equus. In simple terms, a pony is a small horse. Their height is traditionally measured in hands, and a hand is 4 inches (10cm)—said to be the average width of a man's hand. Height is measured from the highest point of the withers, the bony rise at the base of the neck, to the ground.

This unit of height and its fractions are traditionally expressed as, for example, 13hh (13 hands high, or 52 inches) or 15.3hh (15 hands and 3 inches high, or 63 inches) Traditionally, the dividing height limit between ponies and horses is 14.2hh. Just to be different, the tiny Shetland pony has always been measured in inches rather than hands and inches—and some breeds, such as the Arabian and Icelandic, are always called horses no matter what their size.

If you're beginning to think you need a degree in mathematics before you go near a horse, you'll find that some breed societies and competition disciplines which impose height limits now opt to use centimetres. They have done this by rounding up the traditional dividing lines—for instance, 12.2hh, 13.2hh, and 14.2hh—to the nearest centimetre, which in these cases gives 50 inches (128cm), 54 inches (138cm), and 58 inches (148cm).

One word you'll hear frequently when talking about horses is conformation. This means the way a horse is built: a horse or pony with good conformation is one who meets standards that are recognized as being helpful for soundness and athleticism. Although many parts of the horse's body correspond to parts of the human body, they may be given different names.

To understand the characteristics of various breeds and types and how they have evolved and to appreciate how a horse's conformation makes it easier or more difficult for him to do a job, you need to know the names of different parts of the body. These are known collectively as the points of the horse. Some are self explanatory, but a few may sound as if you are entering a secret society. This will help

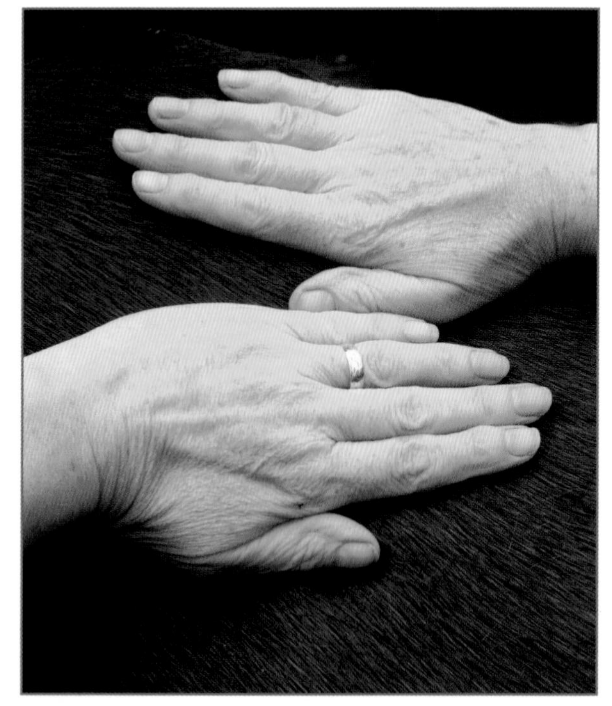

The unit for a horse's height was originally a hand's width.

Measurement is taken from the highest point of the withers.

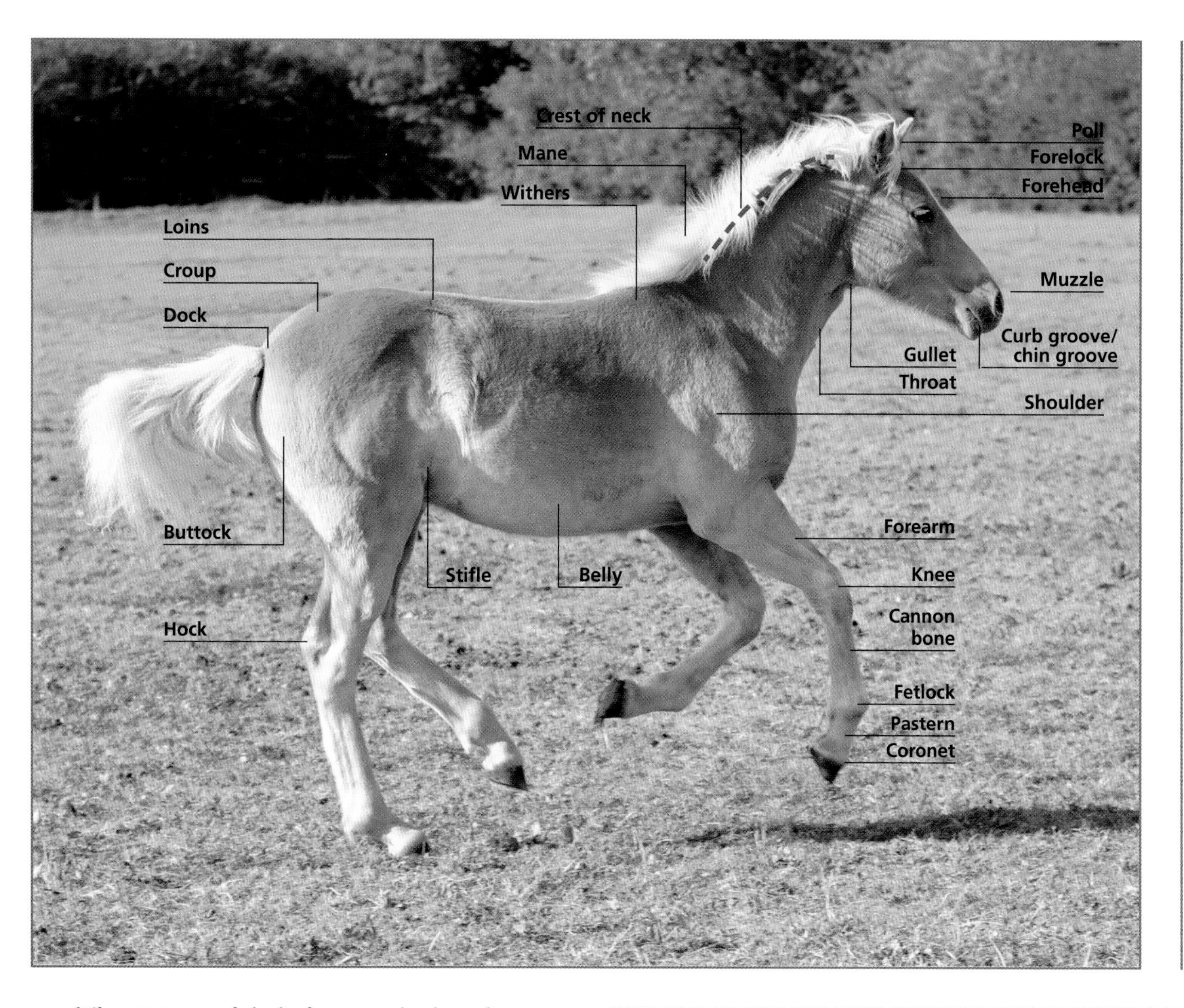

you differentiate your fetlocks from your hocks and your forearms from your cannon bones!

You may hear a horse described as a coldblood, hotblood, or warmblood. Coldblood animals are those belonging to the heavy/draught horse breeds. Hotblood is a term once frequently given to Thoroughbreds and Arabians. It isn't heard so often now, but helps you work out that a warmblood is basically a horse whose breeding combines a mixture of these influences.

It's also important to recognize gender differences. A stallion is an uncastrated male and though many have lovely natures, they are not suitable for novice owners and riders simply because they retain their breeding instincts. Geldings—castrated males—are easier to keep. An uncastrated male up to three years of age is a colt.

A female up the age of three years is a filly and after that, she becomes a mare. Mares, like stallions, retain all their instincts and need owners and riders who understand and appreciate them.

The old saying "Tell a gelding, ask a mare and discuss it with a stallion" has a lot of truth in it—though if you treat any horse fairly, he or she will hopefully respond.

Mares retain their instincts, including maternal ones.

How it all began

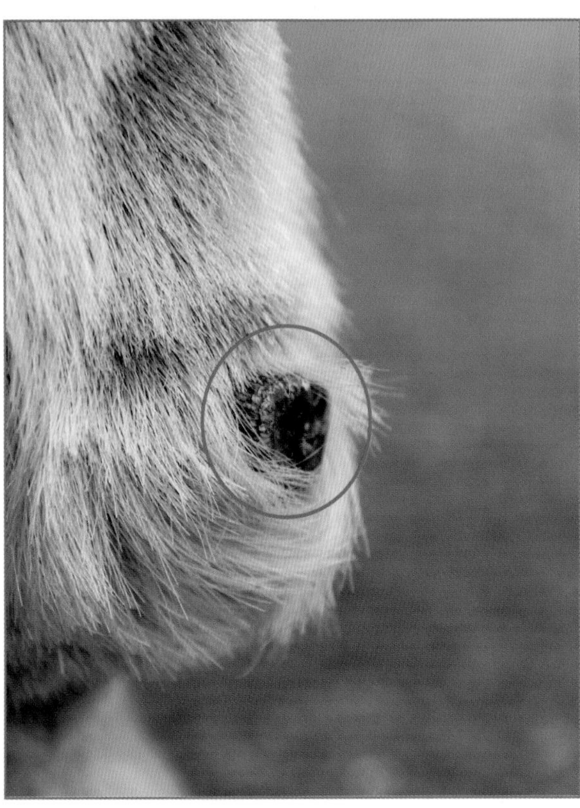
Ergots are vestigial remains of a toe.

When we look at the huge range of breeds and types of horses and ponies, it's hard to believe that they all descend from a creature the size of an average dog who roamed the Earth about 55 million years ago. But from the tiny Shetland pony to the massive Shire horse, they all go back to Hyracatherium (later called Eohippus).

Eohippus actually looked far more like a dog than the modern horse. He had four toes on each foot rather than hooves, a small head, and teeth which were ideal for browsing on leaves and fruits: at that time, grasses had not evolved. Ergots, the calluses underneath the fetlocks, are thought to be the vestigial remains of one of those toes.

As the landscape changed, so did the early horse. Over millions of years, he evolved into a single-toed grazing animal the size of a small pony. A prey animal rather than a hunter, he survived by running away from his enemies rather than fighting.

It's now generally agreed that horses were first domesticated and ridden—as opposed to being kept for food—in Kazakhstan, as early as 3,500 BC. By that time, horses were the size of today's large ponies and their feet had developed into hooves.

Today, there are more than 70 horse breeds throughout the world. The nearest relations to their early ancestors is the Przewalski's horse, which originated in Mongolia and the Konig pony from Poland. Herds of Konigs have been

Przewalski's Horse

introduced to nature reserves to maintain the environment. A few surplus geldings (castrated males) have found their way on to the open market as riding and driving animals, but their build means they are not particularly athletic.

The Fjord pony from Norway and the Exmoor pony from the UK also show physical traits reminiscent of early domesticated horses, but have been selectively bred for modern requirements. This makes their conformation, or shape, more suitable to carry a rider.

Top: Konig pony **Middle:** Fjord horse **Bottom:** Exmoor pony

How breeds developed

Many animals are a mixture of breeds or of unknown breeding, but there are some breeds that have had huge influence on just about all of the others. There are also many which have become popular worldwide because of their characteristics and their ability in various spheres.

The two most influential breeds are the Arabian, or Arab, and the Thoroughbred. The Arabian is said to date back to around 3,000 BC and many people regard it as the most beautiful horse in the world, with its dished face, high head and tail carriage, and overall elegance.

Intelligent and with huge reserves of stamina, it excels in sports such as endurance riding, where riders travel long distances over varied terrain. The Muslim conquests of the seventh century brought the Arabian horse into Christian Europe and it has been influential on just about every existing breed.

The Thoroughbred is the fastest of all the breeds and the world's racing industry developed around it. All Thoroughbreds descend from three stallions brought to England between 1696 and 1728: the Byerley Turk, the Godolphin Arabian, and Darley Arabian. Most authorities believe that the Byerley Turk was, like the other two foundation stallions, an Arabian—though some argue that he was more likely to be an Akhal-Teke, a breed which originated in Iran.

Left: Arabian endurance horse.
Below: Thoroughbred racehorse

Major horse breeds

Over the years, horses have been selectively bred to excel in different jobs. We've already looked at the Arabian and the Thoroughbred and their enormous practical value as individuals and to other world's horses. A minority of Thoroughbreds also has a huge financial value and those which are successful on the racecourse and then retire to stud may sell for millions.

Crossing a Thoroughbred or an Arabian with a different breed or a horse of unknown breeding usually imparts quality to the offspring. Responsible breeding is vital. However fond of them we are, horses are not pets and need to be able to do a job. Breeding from inferior animals means you are likely to get inferior offspring who are unlikely to stay sound and perform well.

It would take a whole book to describe all the horse and pony breeds, so here are some of the ones you are likely to meet—together with a few that are less common and might make you wonder "What's that?" They are in alphabetical order rather than order of popularity and some, such as heavy horses and warmbloods, have been categorized here in groups rather than as individual breeds.

Appaloosa

The Appaloosa is one of the most famous and popular breeds throughout the world because of its spotted coat patterns, but is also believed to be the first American breed to be developed for specific traits. We know that spotted horses were found 20,000 years ago in what is now France because of their depiction in cave paintings, but their real development is traced to the Native American Indians in the early eighteenth century.

Horses were introduced to North America by the Spanish and were traded by many of the tribes. The Nez Perce tribes were renowned as great horsemen and prized animals with spotted coats; it is believed that they were the first to breed them selectively for speed and trainability. They kept the best and traded the not so good stock with other tribes.

When white settlers came to the Palouse region in the Northwest, they called the spotted horses they saw Palouse horses. Eventually, the term "a Palouse" became Appaloosa.

Today, the Appaloosa is popular worldwide. Types vary from those with elegant, sport horse conformation to more solid animals. There are several recognized coat patterns,

including Leopard, Blanket, Snowflake, Snowcap, Marble, and Fewspot. A white ring or sclera surrounds the eye, hooves are striped, and there is pink and black mottled skin around the lips, muzzle, nostrils or eyes. Occasionally, foals are born solid colored and remain so as they mature.

Although the Appaloosa is a breed and not a coat color, it may pass on spot patterns when crossed with another breed.

Above: Pink and black mottled skin around the nose, lips, and muzzle and striped hooves are characteristics of the Appaloosa.

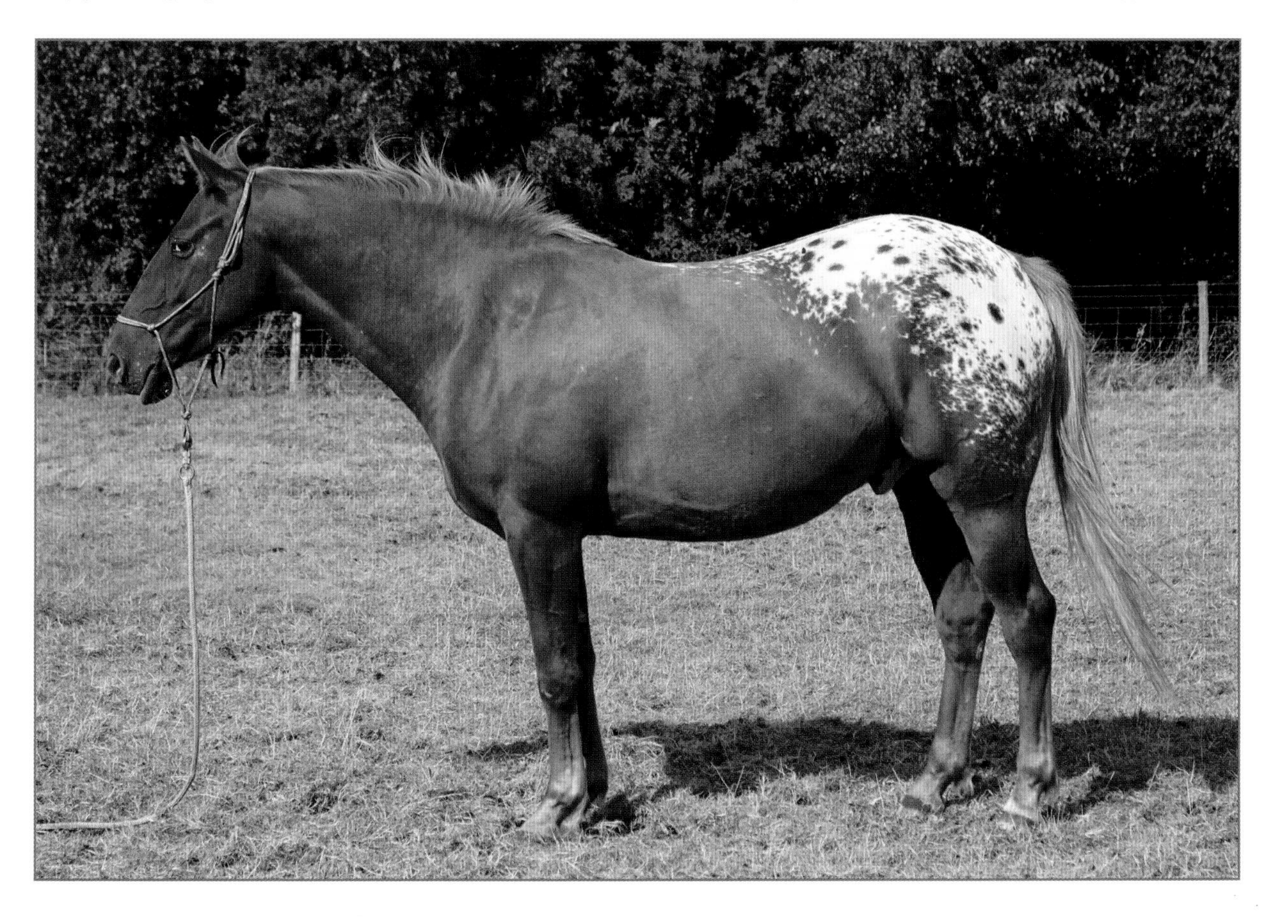

Heavy Horses

The term heavy horse is often used as a generic one for the draft breeds bred to work the land. Draft is a term which describes horses which pull heavy loads; confusingly, the Irish Draught was originally a lighter weight draft horse, which is why it is clearer to label the main heavy horse breeds—the Clydesdale, Shire, Suffolk Punch, Percheron, Boulonnais, and Brabant, to name the best known—as such. They are often characterized as gentle giants, standing over 17hh and weighing more than 1 ton (1,000kg). Many have long, silky hair on their legs, which is known as feathers

The Suffolk Punch is one of the most interesting breeds of heavy horse, because it is always chestnut in color; shades vary, but bright ginger is the commonest. Although no one knows why, in the case of this breed it is spelled chesnut, without the "t." All Suffolks descend from one stallion, which was born in 1768 in Suffolk, England. "Punch" is an old word for a good, stout fellow, which sums up the breed admirably.

Irish Draught

This breed goes back to the days when the horses on small Irish farms were expected to do everything from pull a cart to carry the farmer out hunting. Today, purebred Irish Draughts are becoming increasingly popular again as riding horses and those which are crossed with other breeds—in particular, the Thoroughbred—are successful in all disciplines, especially show jumping.

Horses which are three quarters or seven eighths Thoroughbred, with the remainder of their breeding being Irish Draught, often excel in eventing. Heights range from 15hh to more than 17hh. Good examples of the smaller Irish Draught may be shown as cobs.

Lusitano and Purebred Spanish Horse

These breeds have such a similar heritage that even experts may find it hard to distinguish between them in terms of appearance. In simple terms, the Lusitano comes from Portugal and the Pura Raza Espanola or purebred Spanish horse, is part of Spain's history and culture.

Both are bred and used for a variety of jobs, from herding cattle to performing in the bullring. Today, they are finding favor in dressage and their flamboyance and beauty means many people are attracted to them. Powerful yet at the same time elegant, their average height is around 15.2hh, though demands from today's riders for a taller horse mean you will see many of 16hh-plus.

Morgan

In 1789, a small colt known as Figure was born in Massachusetts. Two years later, he was bought by a schoolteacher called Justin Morgan and taken to Vermont. Over the next 30 years, the little stallion—only just over 14hh—excelled in everything from racing to driving and became known as the "Justin Morgan horse."

Whatever mare he was put to, his offspring inherited his quality, hardiness, and wonderful temperament. Today, Morgan horses retain those characteristics and are popular in the show ring, as performance animals, and in driving. They are renowned for their soundness and longevity, with many living until their thirties.

Quarter Horse

This is believed by some historians to be the first pure American breed and is said to be the fastest horse in the world over a distance of a quarter of a mile—hence its name. Like the Appaloosa, the Quarter Horse descends from Spanish stock and was used primarily for herding and cutting cattle.

Over the years, he has become a pleasure and sport horse throughout the world in Western riding disciplines. He usually stands between 14.2hh and 15.2hh.

Saddlebred

The American Saddlebred is a refined, high-stepping horse which has the ability to perform two extra gaits as well as the walk, trot, and canter. The slow gait is highly collected, four-beat gait and the rack is a higher, much faster version. Horses have been timed racking at 25 mph—which, when you consider that a Thoroughbred racehorse's top speed is around 30 mph, is remarkable.

It was developed by Kentucky plantation owners as the ultimate comfortable riding horse and was first mentioned officially in 1776. The breed became famous during the American Civil War of 1861—1865 and famous generals' horses became as well-known as their owners—in particular, Lee's Traveller.

Although the Saddlebred is perhaps thought of first and foremost as a show horse, he and his partbreds are incredibly versatile and can be as successful as sport and pleasure horses as in the show ring.

Warmblood

Warmblood horses are the success story of modern competition horse breeding. There are many different breeds which come into this category, such as the Hanoverian, KWPN (Dutch warmblood), Selle Francais (French warmblood), and Belgian and Danish warmbloods, but there are many shared breeding lines. Through structured breeding programs and performance testing, breeders have identified lines that combine athletic ability with trainable temperaments. Today, they rule the world at top level in dressage and show jumping, and those with a high percentage of Thoroughbred blood are also successful in the top ranks of eventing.

Wild Horses

Although some breeds are often referred to as wild horses, in practice Przewalski's Horse is the only truly wild breed. However, there are other feral breeds—horses which range freely over wide territories but are descended from domesticated animals. Of these, the best known are the Chincoteague pony and the Mustang.

The Chincoteague pony's heritage goes back to horses which lived on Assateague Island, off the coast of Maryland and Virginia. A popular story is that they swam ashore from a wrecked Spanish ship, but a more prosaic and likely explanation is that their owners grazed them there to avoid fencing and taxation laws.

The breed averages 12-13hh and is very tough and hardy. There are two main herds, which are kept separated, and the size of each herd is restricted to protect the island's ecology. In Virginia, this is done through an annual "pony swim" and auction which has become a tourist attraction. Ponies are swum across to the mainland and many of the foals are auctioned with a view to their eventually becoming riding ponies.

In Maryland, a contraception program involving the use of dart guns began in 1994 and the authorities report that it has proved successful.

True to Type

A horse can conform to a type as well as to a breed and can also be a good example of a type even when his breeding is unknown. The most common types are derived from English usage.

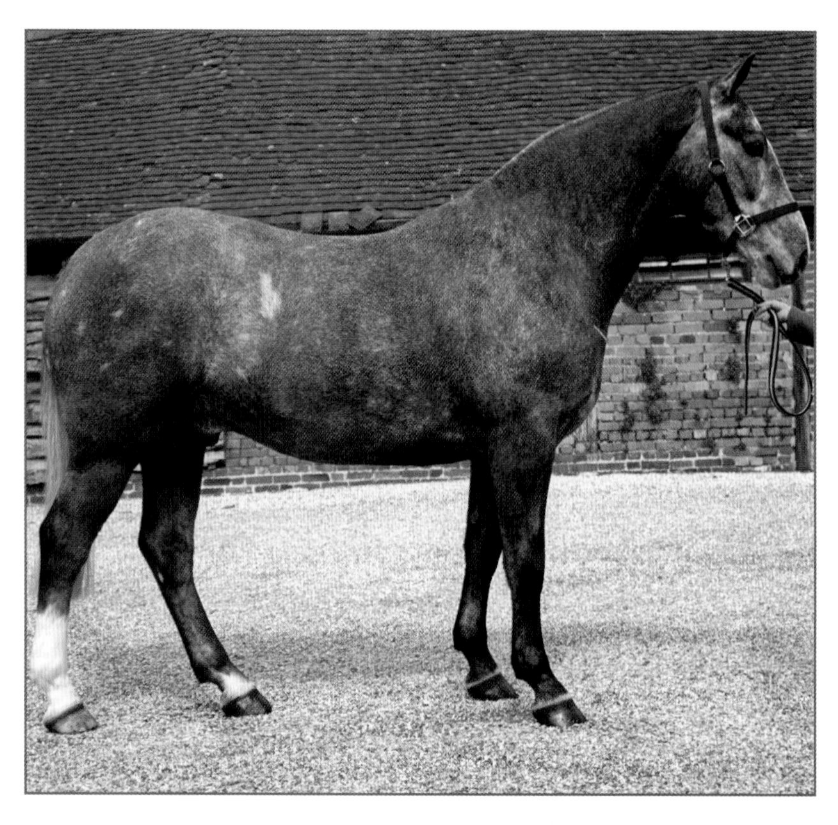

Cob

A cob is a deep bodied, chunky little horse of around 14.3hh—15.2hh who combines workmanlike good looks with the ability to carry weight. A cob who is to be successful in the show ring should, according to tradition, "combine the head of a Duchess with the backside of a cook!" That powerful backside will often mean that a cob is a good jumper. They should have powerful necks and are often presented with their manes hogged (clipped off) to accentuate this.

Hack

In theory, any horse who is used for hacking/trail riding can be called a hack, but the term originates from the days when ladies and gentlemen rode in London's Rotten Row. The park hack was an elegant, graceful animal on which to be seen—and probably show off! The modern show hack carries on that philosophy.

Hunter

Again, any horse who is used for any form of hunting, be it following a trail scent or a prey, can be called a hunter. A "true" hunter is a powerful, well-mannered horse who can gallop and jump and carry the appropriate amount of weight—depending on whether he is categorized as a lightweight, middleweight, or heavyweight—all day.

Riding Horse

This rather vague term is applied to a horse who is in many ways a cross between a hack and a hunter. He should have good conformation and manners and be comfortable to ride at all paces for long periods, without the element of endurance riding.

Major pony breeds

Most of the best known and established breeds of pony come from the UK and Ireland, with the notable exception of the Fjord and the Haflinger. Many breeds have been exported worldwide and you"ll find them being ridden, driven, bred and enjoyed by enthusiasts. They are often referred to as native or Mountain and Moorland ponies.

Don't make the mistake of thinking that because of their relatively small height, ponies can only be ridden by children. Whilst they take on this role extremely well, many of the larger breeds can easily carry adults and you'll see them excelling in all disciplines as well as being owned for pleasure. Because they are so versatile and are cheaper to keep than, say, a large Thoroughbred horse, native ponies make wonderful family all-arounders.

When crossed with the Thoroughbred, the larger native breeds usually produce small horses that combine the best of both worlds and are successful in all spheres. A second cross, producing an animal that is three-quarter Thoroughbred and a quarter native pony, often has the speed and stamina to excel in eventing with the agility and good temperament of the pony.

Native breeds have their own characteristics, such as the black and white hairs of the Fjord horse's mane.

Most breeds have height limits or guidelines. These are referred to in the following pages in either hands or centimeters, according to the governing breed societies' regulations.

Connemara

Over the past few years, the Connemara has become arguably the most popular of the native breeds. As the name suggests, the first recorded breeding of these ponies took place in Connemara, in the west of Ireland. Ranging in height from just over 12.2hh to just over 14.2hh or 50 inches to 58 inches (128cm to 148cm) the Connemara is attractive, athletic, and often jumps well. Connemaras are often grey, though you'll also find bay, brown, black, roan, and occasionally palomino ponies. In recent years there has been a tendency to some Connemaras to grow over the breed societies' height limit of 58 inches (148cm). This has caused a lot of controversy, as although the "over height" animals are popular with many adult riders, experts say it is important to retain the true breed characteristics.

Dales and Fell

Dales and Fell ponies are very similar in appearance, type, and heritage, though they are two distinct breeds. Dales ponies were first used in mines and as farm animals, but make great riding ponies and can easily carry large adults. In wartime, they were used as pack animals and reputedly carried loads of up to 280 pounds (127kg). They are usually black, bay or brown, and less commonly, grey. The Dales is usually about 14hh—14.2hh but there is no upper height limit.

The Fell pony is often slightly smaller and the governing breed society in the UK has set an upper height limit of 14hh. The colors are the same and both the Fell and the Dales have hard, strong feet; they can often be worked unshod. Both breeds have silky feathers on their heels. If you want to pick a real difference, it is perhaps that the Fell tends to have a smaller head in proportion to its body, with a tapering muzzle.

Dartmoor

This breed makes a wonderful child's first pony, as it is small—not exceeding 50 inches (127cm)—not too wide in the body and with comfortable paces. Dartmoors are usually bay, brown, or black, though you also find the occasional grey pony.

Exmoor

The Exmoor is the oldest of Great Britain's native breeds and resembles the earliest domesticated horses. He evolved to withstand harsh temperatures and retains special characteristics that help him do so: his eyes are hooded, called "toad eyes" and his head is proportionately quite long, to accommodate nasal passages that allow air to be warmed as it is inhaled. The Exmoor also has an "ice tail"—an extra layer of hair at the top of the tail which grows in a fan shape and protects him from wind, rain, and snow. Although his height should not exceed 12.3hh, he is easily able to carry an adult.

Fjord

The Fjord pony originates from Scandinavia but in recent years has been exported throughout the world. In the United States and Europe, he has become a popular driving and riding animal. He bears a strong resemblance to primitive horses, with his wide head, small ears, thickset body, and colouring—dun, with an eel (dorsal) stripe running from the forelock, along the back down the tail. The Fjord was the mount of the Vikings, who took him when they raided the Scottish Western Isles: not surprisingly, his influence can be seen on the Highland pony. He is usually 13hh to 14hh.

Haflinger

Bred to work on the steep mountain slopes of Austria, the Haflinger has also become a popular riding and driving pony. He is always chestnut with a flaxen mane and tail or palomino; this, together with his intelligent head, makes him very attractive. The usual height is from 54 inches to 59 inches (138cs to 150cm).

Highland

A great weight carrier, the Highland was traditionally bred to carry shot stags in the Highlands, a dead weight of up to about 195 pounds (89kg). He usually stands between about 13.2hh and 14.2hh and although most Highlands are grey or a shade of dun, there are occasional bays and liver chestnuts.

New Forest

This breed makes a lovely all-round riding pony for both children and adults. The earliest record of horses in its place of origin, the New Forest, England, was 1016 when the people living in what was then a royal hunting ground were granted rights of pasture. Since then, many outside breeds have been introduced, including Welsh, Thoroughbred, and Arab, and later on, Dales, Fell, Highland, Dartmoor, and Exmoor—so it could be said that the New Forest pony epitomises the all-round native pony. Heights range from about 50 inches to the upper height limit of 58 inches (128cm to 148cm) and to be registered with the breed societies, ponies can be any color except piebald, skewbald, spotted, or blue-eyed cream.

Pony of the Americas

The Pony of the Americas is probably the "newest" recognized breed and was officially founded in 1954. The foundation stallion was Black Hand, who was by a Shetland stallion out of an Appaloosa mare.

The breed has been developed as one combining a refined head and overall quality with strong hindquarters. The Pony of the Americas is above all a children's mount and stands between 46 inches to 56 inches tall (11.2hh to 14hh.) Coat patterns vary, but always show a degree of spotting.

Shetland

Everyone knows the Shetland pony, but cute though he may seem because of his small size, he must be treated with the same fairness and firmness as any other horse or pony and needs to do a job in life rather than be a four-legged lawn mower. Registered ponies should not exceed 42 inches (107cm) at maturity and may be any color except spotted. Bred to live in a harsh climate, the Shetland has a double coat in winter which keeps him dry in the harshest weather. He is a lovely riding pony for small children and can also be driven.

Welsh Breeds

There are four distinct Welsh breeds—the Section A, or Welsh Mountain pony; the Section B, or Welsh pony; the Section C, or Welsh pony of Cob type; and the Section D, or Welsh Cob. This is where it can get a little confusing, as while a cob is a type of horse rather than a breed, the Welsh Cob is a distinct breed and as such is dignified with a capital C.

All are tough, hardy, and generally have great presence—a term used to describe the "look at me" quality that some animals (and people) possess. Thoroughbred and Arab influence on the Welsh breeds shows both in this charisma and in their attractive heads. The Section A has an upper height limit of 12hh, the Section B and Section C, 13.2hh, although the Section C is stockier in build. The Section D, shown here, has no upper height limit, but is over 13.2hh.

Breeding—What Breeding?

There are many lovely horses and ponies throughout the world who have no recorded breeding. However, it's often possible to guess at their heritage through studying their conformation. It's important to preserve the gene pool of breeds and in the competition horse world, breeding for performance has become a real science—but there will probably always be animals whose breeding is only recorded for one or two generations, or even not recorded at all.

If a horse has the right conformation and movement to help him stand up to work, as is explained later in this section, and has the right temperament, it may not matter to you where his or her parents come from. In the case of a child's pony, temperament may be the most important thing of all.

Make and shape

Horses, like people, come in all shapes and sizes and to those who have little or nothing to do with them, distinguishing different breeds and types may not be as easy as, say, identifying different breeds of dogs. For instance, while most people could immediately tell the difference between a Springer Spaniel and a Basset Hound, even though they may be of a similar color, they wouldn't necessarily identify which of two ponies was a Connemara and which was a New Forest.

Recognizing differences in conformation (shape and proportions) and learning what gives a blueprint for biomechanical efficiency is a great skill to acquire. Colloquially, it's known as "having an eye for a horse" and while you can try and learn it from books such as this one, you also need to try and look at every horse and pony you meet with a subjective eye. It's a fascinating subject, not least because there is no such thing as a perfect horse and you need to learn how to weigh up plus and minus points.

So why is conformation important? It isn't just a matter of aesthetics, even in the show ring. If the structure of a horse's legs mean they won't support his weight with maximum efficiency, or they affect the way he moves so that he knocks one limb against the other, he is unlikely to stay sound for as long as one whose limbs are better made.

Similarly, a horse can have conformation that makes him well suited to one job, but not as ideal for another. For instance, the angle of a riding horse's shoulder needs to be more sloping than that of a heavy horse used for pulling heavy loads. In simple terms, a sloping shoulder is usually associated with smoother movement, making the horse more comfortable to ride. A steeper angle—which gives what is called a straight shoulder—gives more pulling power, but the horse's legs will move in a more "up and down" way and this may make the ride he gives more jarring.

A sloping shoulder makes a horse comfortable to ride.

Having said that, there are plenty of animals who do a great job even though they don't fit the ideal blueprint. There are also horses and ponies who tick every box in the conformation stakes but don't have suitable temperaments. How much temperament is down to nature and how much to nurture—the way a horse is brought up and trained—is open to debate, but it shows how and why being able to assess a horse is both fascinating and important.

Head to Toe

While people with a good understanding of conformation and a lot of experience can assess a horse when it's fetlock deep in a muddy field and won't stand still long enough to be caught, the easiest and most reliable way is to look at him when he's standing on level ground. When show judges inspect a horse, they ask for him to be positioned so that all four legs are visible.

Before you look at the fine detail of a horse's conformation, see if the overall impression is of a harmonious

The steeper angle of a heavy horse's shoulder gives pulling power.

Above: This Welsh Cob has a front and back end to match. Below: This pony has a weak back end. Bottom: A horse with excellent conformation.

horse. With practice, you'll get a feel for this, but if you're not sure, ask yourself a few questions.

- Does the front end of the horse match the back end, or does he look like the halves of two different animals joined together?
- Do the lines of his neck and body flow into one another, or are there sharp angles?
- Does his head look in proportion to his body, or is it too big or, more rarely, too small?
- Do his legs look in proportion to his body? In particular, you don't want to see spindly legs under a proportionately heavy body, as they have to support a lot of weight.

Squaring It Up

In theory, you can assess a horse's proportions by mathematics. I'm not suggesting you get out a tape measure, but it gives you an idea of what you're looking for. Drawing on suitable photographs in magazines and measuring the proportions can be interesting; you'll find that most successful top class competition horses square up well, but there is always an odd one who breaks the rules.

Ideally, the length of the head should be the same as the length of the neck, the depth of the girth should be the same as the length of the legs, and the height should be the same as the horse's length from his shoulder to the end of his croup.

Always remember—a good rider and correct schooling can make far more of a horse with average conformation than a poor rider and incorrect work can do with a horse who has textbook shape!

Bit by bit

Now that you've got an overall impression, you can look at the component parts that make up the whole horse. Although most people agree that the feet and limbs are the most important, inevitably, your first impression is of the horse's head. There are two points of view on this: one is that the shape of the head doesn't affect the horse's ability and the other is that if you've got to see him looking over the stable door every day, you need to like what you see!

Beauty is in the eye of the beholder. A pretty head is obviously attractive, but a theoretically plainer one can still be handsome. The only thing that really matters is whether the horse's head is markedly too large for his neck and body, in which case it will be harder for him to balance himself under a rider.

A lot is said about a horse having a "kind" or "generous" eye and you may be told that a horse with a small, "piggy" eye will always be mean. This isn't true: a large eye is more attractive, but a small eye is not indicative of a bad temperament. Likewise, a horse with a white ring (sclera) all the way round one or both eyes isn't necessarily wild—it's the way he's made and he can't help showing the whites of his eyes. In fact, with the Appaloosa, this is a breed trait.

You will sometimes find horses where one or both eyes is blue, sometimes called a walleye. This is especially common in skewbalds and piebalds and can look quite startling, but it has no effect on the horse's eyesight or temperament. Some people find it aesthetically unattractive, but again, handsome is as handsome does.

It's often said that you shouldn't look a gift horse in the mouth—because, of course, you can tell his age from the appearance of his teeth—but mouth conformation can have an effect on the way a horse eats and the bit that will be most comfortable for him.

His upper and lower incisors should meet evenly and he should not be markedly overshot or undershot. The first defect is often called a parrot mouth, because it resembles the shape of a parrot's beak with the upper jaw being longer than the lower one. An undershot jaw is the opposite, when the upper jaw is too short. Both problems mean that his teeth will not grind evenly and in really bad cases, this may affect the way he eats.

While you're looking at and in his mouth, notice whether he has thin or fleshy lips, as this needs to be taken into consideration when fitting and adjusting a bit. Does his tongue fit neatly in his lower jaw, or does he have a fat tongue that bulges out at the sides? This will influence the design of bit mouthpiece that he will find most comfortable.

Telling a horse's age from his teeth, or "ageing a horse" as it is often called, is a reasonably accurate guide but not a detailed or definitive one. Where available, breeding documentation which can be matched to a horse's official identity document is probably the safest way, but the following points may be of interest.

Because horses evolved as grazing animals, their teeth erupt very, very slowly throughout their lives and are worn

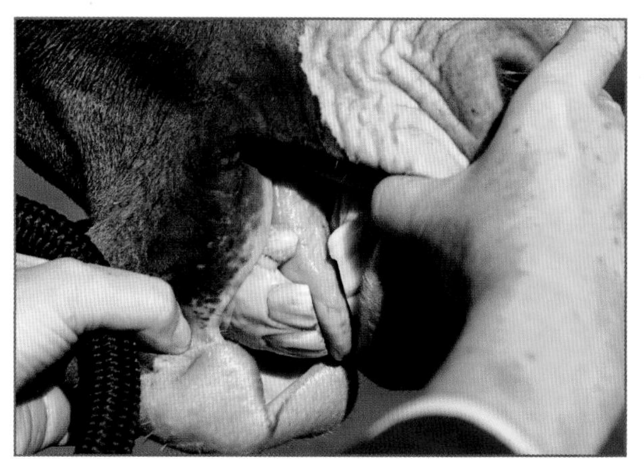

down as they eat—though they also need regular dental care, as explained in Health and First Aid. As the teeth change gradually as the horse gets older, an experienced assessor can get an idea of the animal's age by inspecting them.

It is a process that is most accurate up to the age of seven or eight and after that, becomes much more vague. An adult horse has twenty-four molars and twelve incisors and male horses also have canine teeth or tushes. Many horses also have "wolf teeth," small, shallow rooted vestigial molars. Your vet or equine dental technician may advise that these are removed if they interfere with the action of the bit.

- At one year, the horse has six deciduous or milk teeth in each jaw.
- At two years, he has a complete set of milk teeth which show wear.
- At three years, the two center milk teeth are replaced by permanent ones.
- At four years, the two milk teeth on either side are replaced by permanent ones.
- At four years, the next two milk teeth are replaced by permanent ones.
- At five years, the two corner milk teeth are replaced by permanent ones. The horse is now said to have a "full mouth."
- At six years, the corner incisors are in wear and black marks called dental stars can be seen on the center incisors.
- At seven years, small hooks appear on the top corner incisors.

After this, ageing becomes inaccurate, though there are pointers. In most horses, the hooks on the top corner incisors disappear and at nine/ten years, a line called the Galvayne's groove appears at the top of the corner incisors. This gradually grows down; at fifteen, it has probably reached halfway down the teeth and at twenty, it has probably reached the bottom.

Apart from the fact that the way horses are kept, fed, and ridden can affect the wear on their teeth, there are all sorts of pitfalls for the unwary—so dentition can only be a guide. For instance, unscrupulous traders dealing with the unknowledgeable may try to pass off a two-year-old with a complete set of milk teeth as a five-year-old with a complete set of new adult teeth. Also, at thirteen, a dental hook appears which looks similar to the one on a seven-year-old mouth.

Although the shape of the head is relatively unimportant, the way it is set on to the neck is important. There should be enough space behind the jawbone for the horse to flex easily, so you want to see a nice curve rather than a sharp angle.

Sometimes, people are confused about the shape of a horse's neck because it lacks muscle tone, either because the horse is thin or because he has not been worked correctly. A horse who develops muscle on the underside rather than the top of the neck through incorrect work can be

re-educated, but if ewe necked—which means the neck looks as if it has been set on upside down—the fault will always remain.

Some breeds, notably the Arabian, have naturally arched, graceful necks. Other breeds and types, such as cobs, often have relatively short, thick necks. As long as the rest of the horse is in proportion, it doesn't really matter, but if you get a short bodied horse with an exceptionally long neck or a long bodied horse with a markedly short neck, he will find it harder to maintain his balance when ridden.

Ideally, the neck should join on to well defined withers, as this makes it easier to fit a saddle and for the saddle to stay in place. Many cob type and native ponies have low withers; in fact, their backs can be so flat and wide they resemble table tops! This makes keeping a saddle in place more difficult.

Exceptionally high withers can also make saddle fitting problematical, as it's important that there is enough

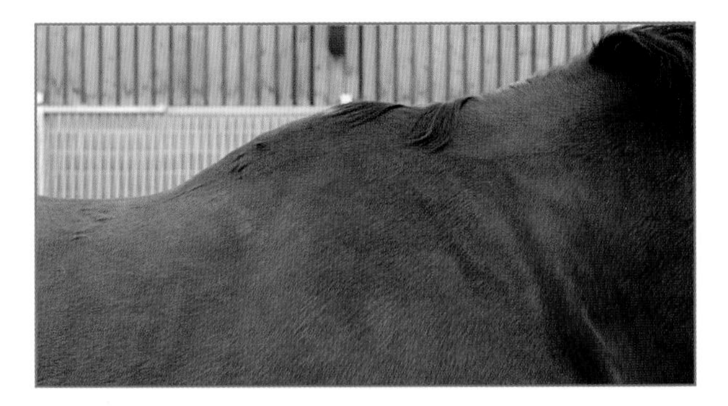

clearance. In this case and that of the horse with low withers, a good saddle fitter will help.

It's often said that a horse should have a kind, generous eye and that a small or "piggy" one can denote a mean temperament. This is unfair, as there are a lot of horses with relatively small eyes who are kind and willing.

As explained at the beginning of this section, the angle of a horse's shoulder will affect the sort of ride he gives you. Compare the angle of the shoulders of the riding horse and the heavy horse and you can see how the first facilitates smooth movement and the second, the pulling power to move a heavy load.

Now you can look at the horse's body shape. His back should be neither too short nor too long, as a long back is a sign of weakness and one that is too short may mean that the horse overreaches (strikes the back of a front foot with the toe of a hind one.) The old dealers' joke is that a horse with an exaggeratedly long back is one that all the family can ride—one behind the other!

Mares tend to be slightly longer in the back than stallions and geldings, as their bodies have to allow room for a foal. There should not be too much room between the last rib and the line of the thigh—as a rough guide, allow an adult's hand span—as otherwise, the horse tends to find it more difficult to put his hindlegs under him and create "pushing power" as he works.

Although some horses have lighter frames than others, you want to see a reasonable depth through the girth. A shallow girth, where the body looks like a narrow tube, leaves less room for the heart and lungs, which may affect stamina when or if the horse is working hard.

In older horses, the back may dip slightly as muscles weaken, though if a horse is working well and his saddle fits properly, there is no reason why he can't retain muscle tone into old age. A roach back, which has a convex outline, is a sign of weakness—though, as always, it depends on the severity of the problem and the job the horse is asked to do.

The part of a horse behind the saddle is his engine, so you want to see a good set of hindquarters: not too short from hip to hock. Ideally, the croup should be broad and fairly flat, though a lot of cobs with draught blood in them have sloping croups; it isn't particularly attractive but doesn't usually cause them any problems.

Look at the horse from the front and from behind as well as from the side. From the front, it's nice to see a chest that is reasonably wide, though not to the extent that the horse has a rolling gait. If the chest is too narrow, it means the horse has "both legs coming out of the same hole" and will probably move badly.

From behind, the points of the pelvis should be of equal height when the horse stands square. If one side is lower than the other, which is often accompanied by muscle wastage, it could be a sign that damage to the pelvis area has occurred at some time. Similarly, is one thigh or shoulder more muscled than its partner? It need not necessarily indicate a problem, but if you were thinking of buying the horse, it would be something to discuss with the vet carrying out a pre-purchase examination.

Although you can forgive a lot of conformation faults—unless, of course, you are looking at a show animal—you can't afford to be too lenient about feet and limbs. The old saying, "No foot, no horse" is as true today as it ever was and the feet and limbs have to support a huge amount of weight.

The front and hind feet should be two matching pairs; you don't want to see one foot noticeably different in size and shape to its partner. However, hind feet are usually slightly smaller than front feet. Sometimes, it's difficult for anyone other than an expert to tell the difference between bad feet and a horse that needs the attention of a farrier, but in general the size of the foot should reflect the size of the horse: you don't want to see soup plates on a small pony or pony feet on a heavy cob.

When you pick up the foot, the sole should be slightly concave, to give a good bearing surface. Horses, like people, sometimes have flat feet and are prone to corns and bruising. Foot conformation is a fascinating topic in itself, so if you get the chance, ask a vet or farrier to explain what you should be looking for.

Well-made limbs are important if a horse is to stay sound over a long working life, but it's rare to find an animal who ticks all the boxes. Similarly, while it's nice to find a horse with clean (unblemished) legs, a horse who has led an active life may have picked up or developed minor blemishes that

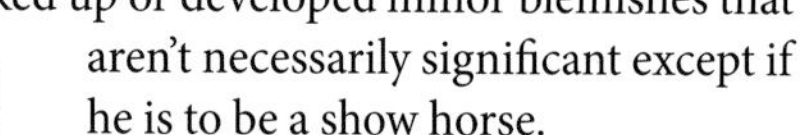

aren't necessarily significant except if he is to be a show horse.

Starting with the front legs, a combination of a long forearm and short cannon bone gives strength and a stride that is long enough for his overall shape. A short stride isn't necessarily a drawback, but a short, up and down stride—often described as a sewing machine trot—isn't comfortable for the rider.

The knee should be wide and flat and when you look at the horse from the front, his cannon bones should come out from the center of the joint, not be offset. A horse who is markedly back or over at the knee is theoretically at greater risk of straining a tendon: back at the knee means there is a concave outline between the knee and the fetlock, and over at the knee means the outline is convex.

Left: Good front legs with long forearms and short cannon bones.

One of the most common blemishes is a splint, a bony growth on the inside of the cannon bone. They may cause discomfort to the horse while they are forming, but, once settled, rarely cause a problem. Judging limb conformation isn't as easy when horses have feather (long hair) on their legs as when the hair is naturally short or has been trimmed, but don't let an abundance of hair trick your eye away from the basic shape.

If you want to estimate how much weight a horse can carry, one of the things to take into account—besides his overall build—is how much bone he has. Bone is the term given to a measurement taken round the widest part of the cannon bone, just below the knee. It is one of the measurements that is still taken in inches.

A hunter or cob type needs 8 inches to 9 inches (20cm to 22cm) of bone to carry a rider of up to about 180 pounds (82kg) and 9 inches to 9.5 inches (22cm to 24cm) to take up to 200 pounds (92kg). However, a horse with short cannon bones and slightly less than the recommended amount of bone will carry as much or more weight than one with more bone but longer cannon bones.

Arabian horses have more dense bone than other breeds and types. As such, they can carry relatively large weights for their size—though as in all cases, don't be too optimistic!

Hindlegs are part of the horse's "engine," so their conformation is important. The worst kind of hock conformation is sickle hocks, where the leg is in front of a perpendicular line dropped from the hock to the ground. Cow hocks, which turn inwards (as do a cow's, hence the name) are sometimes said to be a sign of weakness, but in practice don't cause problems unless the fault is pronounced. They are a natural conformation in the American Paint Horse.

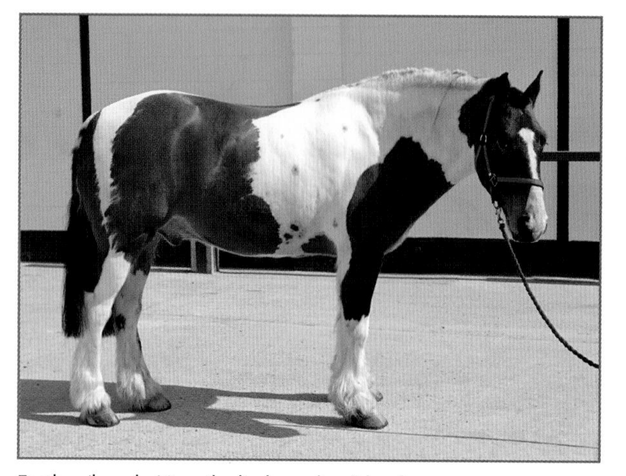

Feather (long hair) on the limbs makes it harder to assess conformation.

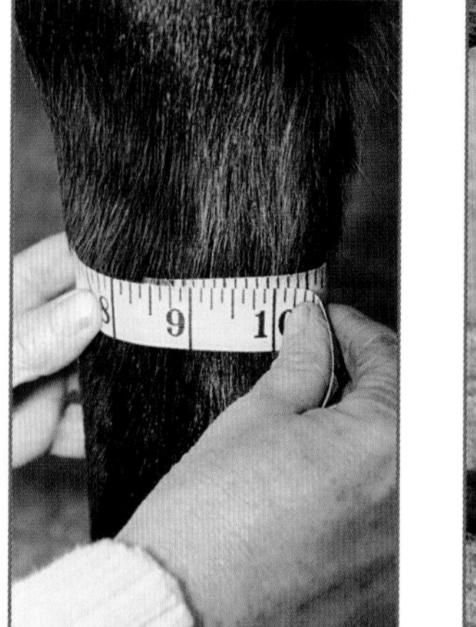

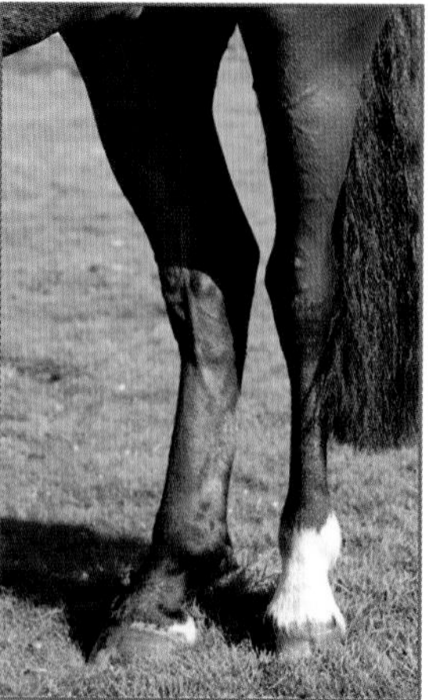

Above left: Bone is still measured in inches. **Above right:** Pronounced cow hocks can be a sign of weakness.

On the move

The way a horse moves is important for practical reasons, not just because it looks good in a show ring. If he moves correctly—straight, as it is often called—without his limbs turning in or out and without striking into one limb with another limb or foot, he will be more likely to stay sound. If his foot or limb deviates from the ideal flight pattern, this puts more strain on the limbs and makes them more susceptible to strain and injury.

However, it's all a matter of degree. There are a lot of lovely horses with less than perfect movement who excel as pleasure and competition animals, such as the cob pictured here. As with everything, it's a case of horses for courses.

The type of movement varies between breeds and types of horses and is linked to conformation and the work he is bred for. As explained earlier, a horse with a more upright shoulder, such as a draft horse, will have greater pulling power and a more upright action than that of a riding horse or racehorse.

To assess the quality of a horse's movement, you need to see him walked and trotted in hand. You would also do this to see if a horse was lame and to try and ascertain which was the affected limb. At top level horse trials and endurance competitions where veterinary inspections are held, horses are trotted up at various stages so that the competition's vets can make sure they are sound.

Ask a helper to walk the horse away from you on level, flat ground, then to turn and walk past and away from you. Repeat the process in trot, positioning yourself so you can watch him from behind, in front, and from either side. Try and look for the following—if you're learning how to do this, it can help to watch a video,

or to watch horses being trotted up for judges in the show ring.

Does the horse touch or knock one leg against its partner on the same side? This is called brushing and can cause injury. In bad cases, a horse may do this every stride, but if his limbs come close together—often called "going close in front" or "going close behind" depending on whether it is for forelegs, hindlegs, or both that are affected—he may only knock himself when he loses balance.

Does he turn one or both front limbs out to the side? This is called dishing and puts strain on the limbs, though is less likely to cause injury than brushing. As always, the potential for problems depends on the severity of the fault: a horse who turns a toe out slightly may never be affected, while one who twists the whole foreleg out from the shoulder is more susceptible.

The opposite scenario is the horse who puts one front foot in front of the other as he moves, so his front feet move on a single track rather than on two. This fault is known as plaiting and it is said that in severe cases, a horse may trip himself up. However, it tends to be most noticeable in the walk and the horse may move more efficiently in the trot and canter.

Sometimes, a horse will clip the back of a front foot with the toe of a hind one. This is known as forging and can lead to tripping and/or shoes being pulled off. It's something that young and/or unbalanced horses may do when ridden until they learn to adjust their balance under the weight of a rider.

Over-reaching is similar. When this happens, the horse strikes into the heel of a front foot with the toe of a back one. This can result in a wound that is sometimes slow to heal; because of its position it opens every time the horse takes a step.

Although it's obviously better for a horse to move correctly, you can do a lot to protect a horse with imperfect movement from injury by fitting protective boots.

Understanding the footfalls in the different gaits will help you assess the way a horse moves as well as improving your riding. The sequences are:

Walk: four-beat rhythm, left hind, left fore, right hind, right fore.

Trot: two-beat rhythm, diagonal pairs of footfalls with a moment of suspension between them. The right hind and left fore touch down together, then there is a moment of suspension as the horse comes off the ground and the left hind and right fore come down together.

Canter: three-beat rhythm and one foreleg, called the leading leg, takes a longer stride than the other. In simple terms, when the horse canters to the left, his left leg is the leading leg because it is easier for him to balance this way. When he is cantering to the right, his right leg is the leading leg. The sequence in right canter is left hind, then right hind and left foreleg together, then right foreleg, followed by a moment when all four feet are off the ground.

Gallop: a faster version of the canter but with a four-beat rhythm. The sequence of footfalls for left gallop is right hind, left hind, right fore, left fore.

Occasionally, you find horses who move in a different way from the norm, either because they have been trained to do so for harness racing or because it is a breed characteristic. Harness racing horses, who travel as fast as possible while staying in trot, move laterally, with the foreleg and hindleg on the same side moving forwards together. The Paso Fino also has lateral movement and is bred to give a comfortable ride.

Colors and markings

There is a huge range of coat colors and markings. Funnily enough, the ones you are least likely to see are, technically speaking, black and white. A black horse is only truly black if he doesn't have a single brown hair, but most have brown hairs on their muzzles, inside their ears or perhaps on the inside of their thighs. Strictly speaking, white horses are always officially grey, though traditionally the Lipizzaners of the Spanish Riding School are often known as the "dancing white horses" of Vienna and Camargue horses are also referred to as white.

Some breed societies refuse to accept particular—usually skewbalds and piebalds, with patches rather than solid coat patterns—into their registers, because they claim this weakens the gene pool. However, there is a lot of argument against this and the situation may eventually change.

There are a lot of myths about coat colors. The favorite is that chestnuts are always hot tempered; the same is said about red-headed people and is patently untrue. It is also said that horses with "weak" colors such as light bay or pale chestnut have weaker constitutions and again, there is no truth in this.

There is an old saying that "A good horse is never a bad color." This really is true—because if a horse is a good one, it doesn't matter what colour he is.

In alphabetical order, the basic coat colors are:

Bay: Reddish brown to dark brown with black points (mane, tail, lower legs, and tips of ears.)

Black: A true black is rare and has no brown hairs anywhere on the face, body, or legs.

Brown: A brown horse has a mix of brown and black hairs in his coat and may have black points. Many have lighter colored muzzles.

Buckskin/Dun: To most eyes, these are too similar to differentiate, though some authorities say a buckskin always has a true black mane and tail. Buckskin is an American term and dun, an English one. This color often incorporates a dorsal eel stripe, a dark line along the back which is a primitive marking seen on animals such as Przewalski's Horse.

Chestnut: There are various shades of chestnut, from a dark gold through to liver chestnut, which has brown hue. A chestnut with a lighter mane and tail is said to have a flaxen mane and tail.

Grey: Ranges from iron grey, a dark grey coat with a sprinkling of white hairs through it, to the classic rocking horse dapple grey. Grey horses nearly always get lighter as they grow older and some become fleabitten. This unattractive term is used to describe an attractive color, where the base coat of white hairs is flecked throughout with brown.

Palomino: The head, limbs, and body are golden and the mane and tail, white. Ideally, the main color should be the shade of a newly minted gold coin.

Piebald: The coat is divided into black and white patches, with no brown hairs. If any brown hairs are present, even on just the muzzle area, the animal is technically skewbald.

Roan: A mix of dark and white hairs. One of the most attractive is strawberry roan, where the mix of chestnut and white takes on a "pink" tone. Bay roan is slightly darker and blue roan has mix of black or brown mixed with white that has a "blue" tinge.

Skewbald: The coat is divided into patches that are white and any other color except black. This can be anything from palomino or dun to chestnut or bay.

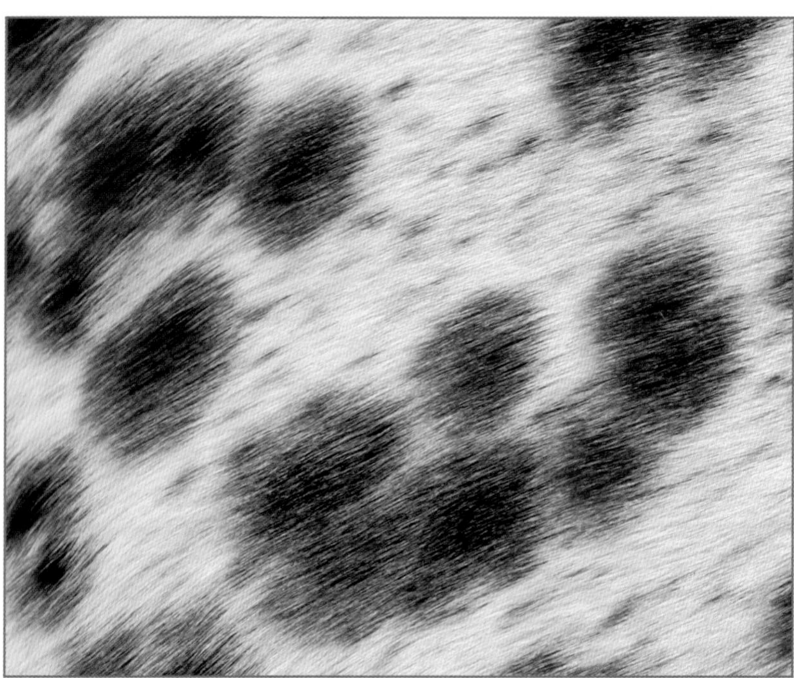

Spotted: Irregular small, dark patches, usually black, on a white/grey base. This is often called Appaloosa coloring, but the Appaloosa is a breed and may have one of several recognized coat patterns.

A star is an irregularly shaped white patch on the forehead.

A broad white blaze is particularly striking.

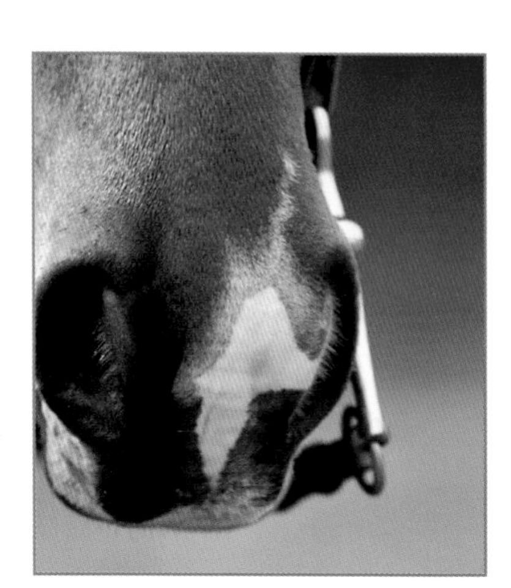
A snip can extend into the nostril, as shown here.

A horse or pony can also be described and identified by markings on his head and limbs and by the placement of whorls, where hairs radiate out from a central point. The placement of whorls is individual to every horse and should be recorded on his ID document. If confronted by two seemingly identical horses, each with an ID document, a knowledgeable person such as a veterinary surgeon would be able to distinguish between them by identifying features such as whorls.

The commonest facial markings are star, stripe (or strip), blaze, and snip, or a combination of them. A star is an irregularly shaped white patch on the center of the forehead. A stripe is, as its name suggests, a narrow white stripe down the center of the face and a blaze is a broader version. A snip is a white marking on the muzzle and may extend into the nostril.

The three main types of leg markings are white socks and stockings and black ermine spots. Socks are white coloring anywhere up to the knee or hock and stockings extend over the knee or hock. Ermine spots can be anywhere around the coronet.

Socks finish below the knee or hock.

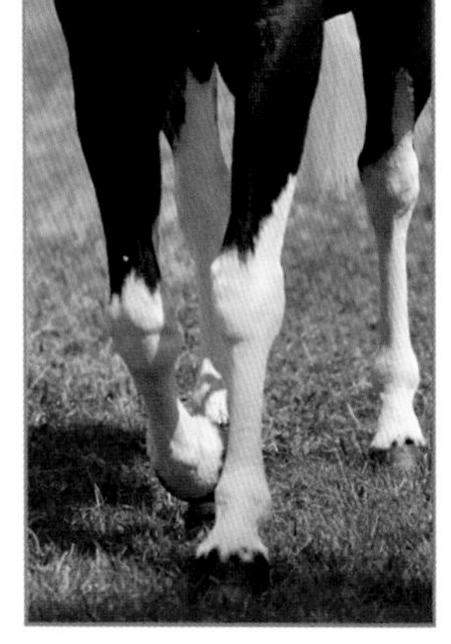
Stocking extend above the knee or hock.

A horse may also have acquired markings, such as brands and freeze marks, permanent blemishes and scars and white hairs that have grown through on the site of an injury or where a badly fitting saddle or rug has applied pressure.

Identification marks

Brands, freeze marks, and tattoos are used to identify horses, either by signifying that they've been inspected and graded as being good examples of a particular breed or as a security measure against theft.

Branding, where marks are burned into the skin with a hot iron, is today controversial and in some cases illegal. Many people believe that advances in microchip technology make it unnecessary. However, there are claims that in some situations—especially when dealing with feral ponies such as those on Exmoor, which are branded routinely as foals—it helps identify a pony more easily.

In freeze marking, irons similar to those used for hot branding are dipped in liquid nitrogen and dry ice, then applied to the horse. This kills the pigment in the coat so that the new hair that grows through is white. When a grey horse is freeze marked, the irons are applied for longer so that there is no hair regrowth and the marks appear as bald areas.

Lip tattoos, where horses are marked with identifying numbers on the underside of the top lip, are used on Thoroughbred horses in some countries.

PART 2

INSIDE THE HORSE'S MIND

To build a good relationship with a horse or pony, whether in the saddle or on the ground, you have to learn to think like one. This doesn't involve any mysticism or "horse whispering," simply appreciating his instincts and the way he reacts physically and mentally.

Horses, like all domesticated animals, retain their instincts. Just as a friendly family dog will retain traits of his wolf ancestors, perhaps by chasing and even catching rabbits, so a horse retains the instincts and natural behavior of his wild antecedents.

A dog is a predator, but a horse is a prey animal. This is the overriding factor in how he relates to his environment, to you, and to anything he is asked to do. A horse's main survival mechanism is flight, not fight—if he is faced by anything that he finds frightening, he will run away, whereas a predator animal may be more likely to stand and fight.

That isn't to say that horses never exhibit aggressive behavior. Obviously they do, as anyone who has been on the receiving end of a kick or bite will testify. However, usually this is not overt aggression but is performed in a defensive way—you will probably never see or hear of a horse attacking a human and though horses can demonstrate aggressive behavior to each other, especially when turned out in a group, this is usually part of establishing a "herd hierarchy." Once the pecking order is decided, the group will usually live in harmony.

It is also important to remember that horses are herd animals and need the companionship of their own kind. It is unfair to keep a horse or pony on his own and although some may seem reasonably happy with companions such as sheep or goats, they are always happiest with other equines.

The nicest natured domestic animals will still retain their instincts.

This may or may not include donkeys: some horses get on well with them, while others view them with great suspicion.

If two or three horses of the same sex are kept together, they will usually, though not always, form a good relationship. In a larger group, horses will often form "pair bonds," rather like finding best friends, within the herd setup.

Opinions vary on whether mares and geldings should be kept together or in same gender groups. Some setups work well, while others result in problems. For instance, sometimes two geldings will get on perfectly well but if a mare is introduced, will become jealous of each other.

Never underestimate girl power. In the wild, a stallion may lead a herd but it is the dominant mare who is the main boss. In domesticity, mares can be just as confident—or bossy—within a group.

Horses are herd animals and need the company of their own kind.

Senses and sensibilities

Horses have finely tuned senses of hearing, smell, sight, and touch, though in a different way from our own senses. For a start, their physical characteristics are different: a horse's eyes are set at the sides of his head rather than in the front and his outer ears not only funnel sound, but can move almost 180 degrees from front to back.

Sight

A horse is at his most vulnerable to predators when grazing, with his head down. Because of the way his eyes are positioned, he can still see nearly all the way round him—and if he moves his head slightly, he can gain a full circle of vision.

A horse's window on the world is different from ours. For a start, he has both monocular vision (which allows him to see different things through each eye) and binocular vision (which means he can focus on things with both eyes at the same time.) He can also switch between the two.

However, because his eyes are situated high up and at the sides of his head, he has small "blind areas" directly in front of and behind him when his neck is straight. This means that when he has the freedom to do so, he will raise or lower his head to see better; you'll notice that when a horse sees something interesting in the distance, he will raise his head.

It also means that though we ask horses to work "on the bit" in the dressage arena and when schooling, which means his head is on the vertical, he needs to be able to adjust his head carriage when jumping or, in the case of the working Western horse, cutting cattle. If the position of his head is too restricted, he won't be able to judge distances correctly.

It used to be thought that it took a long time for horses' eyesight to adjust between areas of darkness and brightness. However, we now know that they can adjust relatively quickly. We also know that although horses do not differentiate between colors in the same way that we do, they are not color blind.

Hearing
A horse's sense of hearing is much sharper than ours. He can not only pick up sounds from a greater distance, but also hear higher and lower frequencies—which may be one reason why a horse will sometimes seem interested in or worried about something when you can't spot the cause.

They are also able to differentiate between similar sounds. Many owners who keep their horses in livery yards (training barns) will know that their horses recognize the sound of their car engine and will whinny, while ignoring the arrival of others. Sadly, they are probably looking forward to the arrival of a meal rather than the arrival of an owner!

Smell
The sense of smell is also stronger in horses than in humans and, as with all the senses, is part of their survival package. Horses breathe through their nostrils, not through their mouths and as well as being mobile, the nostrils lead into long nasal passages.

It is well documented that horses can take a violent dislike to some odors—that of pigs is a classic example. The sense of smell is also part of the breeding instinct; a stallion can smell a mare who is in season from a long way away and individual scents are very important in the bonding process of a mare and foal.

Horses have an extra sensory weapon—the vomeronasal or Jacobson's organ, which is a sac at the top of the nasal passages. They use this to investigate any unusual or particularly stimulating scents by demonstrating behavior known as Flehmen. In this, the horse raises his nose and curls his top lip; it looks as if he is laughing, but in fact he has taken in the intriguing smell and is now partly closing off his nostrils so that the Jacobson's organ can process it more fully.

Taste
Although horses are born to eat and if kept naturally, spend much of their time grazing, they have a discriminating sense of taste. They are fastidious and will usually only eat moldy hay or drink dirty water if they are starving or very thirsty, which is one of many reasons for ensuring that all food is of good quality and that there is a constant supply of clean, fresh water.

Horses like to groom each other with their teeth.

How horses communicate

Although horses have a wide range of vocal sounds, from the soft call of a mare keeping her foal close to her side to the explosive snort of a horse indicating that he's curious or excited, they communicate mainly by body language.

The horse that is the boss of an established group will hold up his head and walk towards an animal he wants to move out of his way, looking straight at him and, if he wants to make his point clearer, laying back his ears and showing his teeth. A young horse that wants to show that he's submissive to an older one, or a mature animal that wants to show that he knows he's lower down in the pecking order, will lower his head and make non-aggressive snapping motions—the gestures a foal makes when he wants to feed from his mother.

Once horses get to know each other they'll happily graze side by side, and really good friends may look as if they're sharing the same blade of grass! However, until they're totally relaxed in another's company, they'll respect each other's personal space.

Horses are rarely aggressive, but will kick and bite if provoked. Even then, they usually make their intentions clear by laying back their ears and raising a hind leg in warning before carrying through their threats.

Young horses, like any young animals, will also play. This usually involves mock fights, with the instigator rearing and boxing with his forelegs to encourage another youngster to join in. Older animals will also play, but are more likely to enjoy themselves by galloping round the field.

Some horses will reject anything that doesn't taste "right," even down to a teaspoon full of medication or dietary supplement mixed in with the usual feed. The only way to get round this is to disguise the taste with something the horse finds pleasant, such as soaked sugar beet pulp or molassed chaff.

Unfortunately, horses may eat poisonous plants, which is why good pasture management is so important.

Touch

Horses are tactile animals and use their whole bodies to react to their environments and to each other. The whiskers round their muzzles are used as feelers and the muzzle and lips are particularly sensitive. While we would touch something with our hands, a horse will explore it with his muzzle.

They can also feel the lightest touch: watch a horse when a fly lands on him and he will twitch his muscles to get rid of it. At the same time, he will use his teeth to "groom" another horse, usually just behind the withers or along the back.

Making Sense of It All

So how does our knowledge of the horse's senses help us? For a start, we need to respect the differences between horses and humans and make sure we are not inadvertently making him uncomfortable or unhappy.

Avoid loud noises. Don't shout, or play a radio loudly on the stable yard. Horses respond to quiet words to soothe them or a slightly sharper reprimand, but never shout.

At the same time, don't wrap up your horse in cotton wool. Experiencing different sights and sounds is all part of his education and life experience and horses have to get used to, say, loudspeakers at a show.

Don't insist on stroking a horse who doesn't want to be fussed over. Some horses like attention all the time, but others prefer their own space. One thing most horses enjoy is being scratched just behind the withers, but watch out that he doesn't try and return the compliment by nibbling at your back too enthusiastically!

Body language

Learning to read a horse's body language and influence it by your own will improve your communication with him. For a start, you'll understand how he is feeling at any given time.

The horse who is healthy but resting or even dozing will stand with his head lowered, his ears relaxed, and often a resting hindleg. If he is really chilled out, his eyes may be half closed and his bottom lip drooping.

An excited horse will have his head raised and his ears pricked forwards. Everything about him will show you that he is alert—he may snort, prance, and/or raise his tail. A horse who is showing aggressive behavior, either towards another horse or to you, will usually lay back his ears and stretch out his nose, threatening to bite or even baring his teeth, and carrying the threat through. Alternatively, he may raise a hindleg and threaten to kick—or actually lash out.

Remember that horses can't talk, so can only tell you how they are feeling through their body language. It's up to you to interpret it correctly and it isn't always as easy as it sounds.

Curious or aggressive? His ears are back, but his eyes and lips show no sign of tension. In fact, this horse was curious about the photographer's camera!

For instance, a horse who is ill will often adopt the posture of a resting horse. Don't panic every time he enjoys a doze in his field or stable, but do be aware of signs of health and ill-health.

Another common scenario is the horse who has been handled inconsiderately and caused discomfort or even pain. For instance, a horse whose saddle does not fit him or who has been roughly girthed up may react by stepping away as you go to put the saddle on, laying back his ears, and swinging his head to nip as you tighten the girth or kicking out.

It takes a long time for a horse to get over bad experiences and the memory may never go—so if you find your horse behaving in this way, don't punish him, but get advice on handling him in a way that is considerate but keeps you safe.

Over the past decade there has been a huge surge of interest in how adjusting your body language can influence your horse. This is something horsemen (a term which also encompasses women!) have always known and used, even if instinctively rather than by analyzing what they do, but some trainers have been able to explain their methods in ways that those with no previous experience of horses can appreciate.

However, it's important to look beyond the glamour and glitz of high profile people (or those who claim to emulate them) and hone in on the things that really matter. It's also important to keep your common sense.

On the road

At some stage you'll have to transport a horse, whether you're bringing home a new purchase, competing, or visiting a veterinary center. This means not only keeping your horse relaxed and comfortable when loading and on the road, but being aware of driving or towing techniques and skills.

Don't assume that because you can drive a car, you can tow a trailer. You need first to make sure to get some expert tuition. Practice maneuvering and changing gear without a horse on board until you can do it smoothly, and be aware that you need to be even more careful when transporting a live cargo that moves.

Inexperienced horses need to be taught to load so that they associate it with a pleasant experience. Again, this should be done with expert help, perhaps using an experienced horse that will load confidently in order to give the inexperienced one more confidence.

Even traveling equipment will seem unfamiliar and perhaps worrying to a horse that has never worn it before, so get him used to wearing protective items such as leg bandages or traveling boots at home until he's confident. When you can lead him round the yard and he stays calm, you can practice loading.

It's sensible to wear a hard hat, gloves, and sensible footwear in case the horse spooks and, for instance, treads on your toes. Make sure the vehicle is light and inviting and the ramp is on stable ground, walk forward confidently, give the horse time to see where he's going, and you'll have the best chance of him loading calmly.

All vehicles should be well-ventilated without being too drafty and should be cleaned out after every journey. Think of your trailer as a stable on wheels and try to provide a healthy environment, as explained in the next chapter.

For instance, the fact that a horse has been educated to the stage that he will accept a saddle, bridle, and rider in an enclosed area doesn't mean he understands the aids in the way that a well schooled horse does, or that he is safe to ride in all circumstances.

The most useful things to appreciate are the use of eye contact and posture. In simple terms, if you square your shoulders, make yourself as "big" as possible, and look a horse in the face, he will be likely to back off. However, if you slump your shoulders, lower your head, and avoid direct eye contact, the horse will be likely to stand still or come towards you.

This can be used in all sorts of ways, from teaching a horse not to invade your space when you are handling him on the ground to catching him. March towards him with a headcollar, staring directly at him and he isn't going to want to come towards you—but approach him from the side, with a casual posture and you'll get a better response.

When we are taught to handle horses, it is traditional to do many things from the near side (left-hand side.) We lead, tack up, and mount from the near side, not for a sensible reason but because of a leftover relic of the past! When men wore swords for military purposes (and even now, for ceremonial ones) they wear them on their right-hand sides. If you don't want to have a nasty accident, you therefore need to handle and mount your horse from the left-hand side.

As a sword is unlikely to be part of the modern rider's attire, it is far more sensible to accustom your horse—and yourself—to tasks being done from either the near side or the off (right-hand) side. Not only will this be useful if you are ever in a position where it's impossible to do something from the near side, your horse won't become "one sided."

In general, be positive but gentle in the way you handle horses and don't set yourself up for failure. The obvious example of the latter is the person who stands at the top of a trailer ramp trying to pull the horse towards him or her—and as horses do anything possible to avoid stepping on people, that approach is a no-brainer from the start.

Make sure the vehicle is light and inviting and the ramp is on stable ground, walk forward confidently, and give the horse time to see where he is going and you'll have a much better chance of success.

How horses learn

How intelligent are horses? That's a question that is hotly debated and the first thing you have to decide is how you are going to define intelligence.

One of the leading researchers into equine intelligence, Dr. Evelyn Hanggi, founder of the Equine Research Foundation in California, says that while for many years horses were thought of by many as having small brains and little intelligence, in reality, they manage both ordinary cognitive tasks and mental challenges.

In the wild, they cope with challenges ranging from predators which change their location and habits to food and water whose distribution and availability varies. The challenge for domesticated horses may be even greater, as they have to live in environments that may be unsuitable and suppress their instincts at the same time as learning tasks that do not conform to their natural behavior.

Horses learn through both positive and negative reinforcement. Through positive reinforcement, a horse performs an action to achieve a reward he finds pleasant; through negative reinforcement, he avoids something that is unpleasant.

Although negative reinforcement is a traditional tool in horse training and in its worst form, involves force or inflicting pain or discomfort, it doesn't have to be bad—applying a gentle leg aid and releasing the moment the horse obeys is one example. Most experts now agree that both positive and negative reinforcement have their roles and that positive reinforcement can be particularly powerful.

However, you have to use it in the right way—and not by mistake. If a horse kicks his stable door when he anticipates the arrival of a feed and you quickly provide it to stop him making a row, you are positively reinforcing unwanted behavior. Instead, try and ignore the banging until the horse stops—and then give him his feed. It's easier said than done, but worth persevering with.

Different techniques can be used to help a horse to learn. One of the simplest is habituation, something that happens to all of us—human and animal—in ordinary life. It simply means that when a horse is repeatedly exposed to something, such as having his headcollar put on or weather conditions such as rain, he becomes used to it.

Desensitization is useful for introducing new experiences and helping hypersensitive horses accept things they find difficult to cope with. For instance, when a young horse is introduced to the idea of wearing a blanket, you don't just throw it on him and hope for the best—you fold it up, let him sniff it, rest it over him, and walk him round, then gradually unfold it and fasten it, and so on.

A similar technique, often called "sacking out," can be used to help spooky horses accept potentially frightening things. Working in a safe environment, fold a piece of plastic sheet until you can hold it the palm of your hand, and work it over the horse's body, legs, and face as if you were grooming him. Once he is happy with this, unfold it, little by little, until you reach the stage where you can move it over his body, legs, and face without worrying him. If at any stage he is worried, go back a step. Eventually—and it will often take several sessions—you will be able to flap the plastic over him without him being worried by it.

Pavlovian conditioning is where a horse is conditioned to give a particular response to a particular word or action. To take the example of the young horse again, you can teach him to go into trot on the lunge by flicking a lunge whip and using a vocal command at the same time; eventually, you will be able to use the vocal command alone.

Operant conditioning is particularly effective with horses. When a horse starts to learn the meaning of a new stimulus—whether it's a word or a flick of a whip—it will offer the desired response sometimes, but not every time. Using positive reinforcement when you get the desired action, perhaps through praise or, when appropriate, food, will help the learning process.

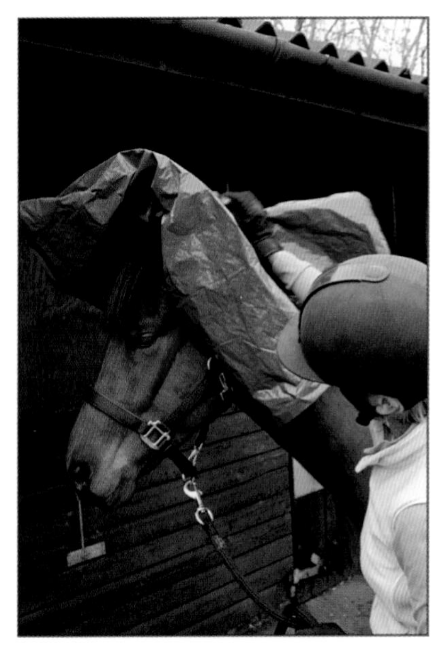

Rewards—and the Reverse

In simple terms, a horse should be rewarded the moment he gives you the response he wants and the opposite applied when he doesn't. Although technically this could be described as rewards and punishments, there are too many unfortunate connotations with the word punishment for it to be appropriate.

You need to be careful about how you define and offer rewards. The obvious answer is a food reward, but while this may be appropriate in some circumstances—such as when you catch a horse in the field—it's not advisable to feed treats constantly. This will probably encourage the horse to invade your space, push at your pockets, or even nip when the desired reward isn't forthcoming. Just as important, he will employ the same tactics when other people are near him.

Instead, a soft word or two of praise, such as "Good" or "Good boy/girl" will be associated by many horses as recognition that they have done what you asked. Remember, too, that simply ceasing the application of an aid once the horse responds is a reward.

If you close your lower legs round your horse to ask him to move forward, release them as soon as he does. If you keep using the leg aid, or even worse, tap your legs on his side in a constant nag, nag, nag, he will tune out and stop listening.

In the same way, if you are leading a horse and apply pressure on the lead rope to ask him to stop, release it as soon as he obeys. If you keep giving a signal after he has performed the desired action, he will be confused as to what response you want and you'll have a communication breakdown.

For some horses, a stroke on the neck or a scratch on the withers is a pleasurable reward. It's often suggested that you should pat a horse on the neck, but many dislike this.

One of the biggest mistakes—and it's easily done—is to anthropomorphize a horse. Just as many dog and cat owners treat them as if they think in human terms, horse owners can tend to think they have the same likes, dislikes, and attitudes as they do.

Horses are not deliberately "naughty" in that they want to annoy you. A horse doesn't come out of his stable and think, "I'm going to wind her up today by refusing to canter on the left lead" or "I don't fancy being ridden, so I'll buck her off."

He may well object to doing something because it causes him discomfort, or he may not understand what you are asking him. Getting angry with him and hitting him won't do any good, but it will make things worse. It's your responsibility to find out why he can't or doesn't want to do something and find a way to make it easier for him, either by correcting your riding, finding a new way to make a task easier for him, removing any discomfort, or all three.

Similarly, you must be consistent in the way you handle, ride, and look after him. It's no good laughing at him one day when he bucks because you're in a good mood and it seems funny, then smacking him the next time because you've had a bad day at work.

Environment matters

While caring for a horse is dealt with in the next section, it's important to appreciate that you can't keep him healthy and happy unless you understand and take into account his mental and physical attributes. For a start, he must have a lifestyle that takes into account the fact that he has evolved to move around in open spaces and to spend more than half his time grazing.

In the wild, a horse will cover many miles in a day searching for suitable grazing. The domesticated horse has a much easier life, as he has food on tap in (hopefully) a well-managed field. In fact, he can often have too much of a good thing, as is explained later in this book. However, he needs to be able to mimic natural behavior for his mental and physical well-being, so it is essential that—apart from in exceptional circumstances, such as when his mobility has to be limited because of injury or illness—he is allowed as much time as possible in the field.

Some people embrace this to the extent that they believe all horses should have a 24/7 outdoor lifestyle, but this isn't always possible and there may be times when it isn't advisable. If shelter is inadequate or land becomes heavily poached, most horses are happy to be kept on a combined system, where they are perhaps turned out during the day and stabled at night.

It can't be emphasized too much that the horse is a herd animal who needs the company of his own kind. Turnout time in a safe, well cared for field with congenial company is essential for his well-being. Horses, like people, need time to relax and socialize.

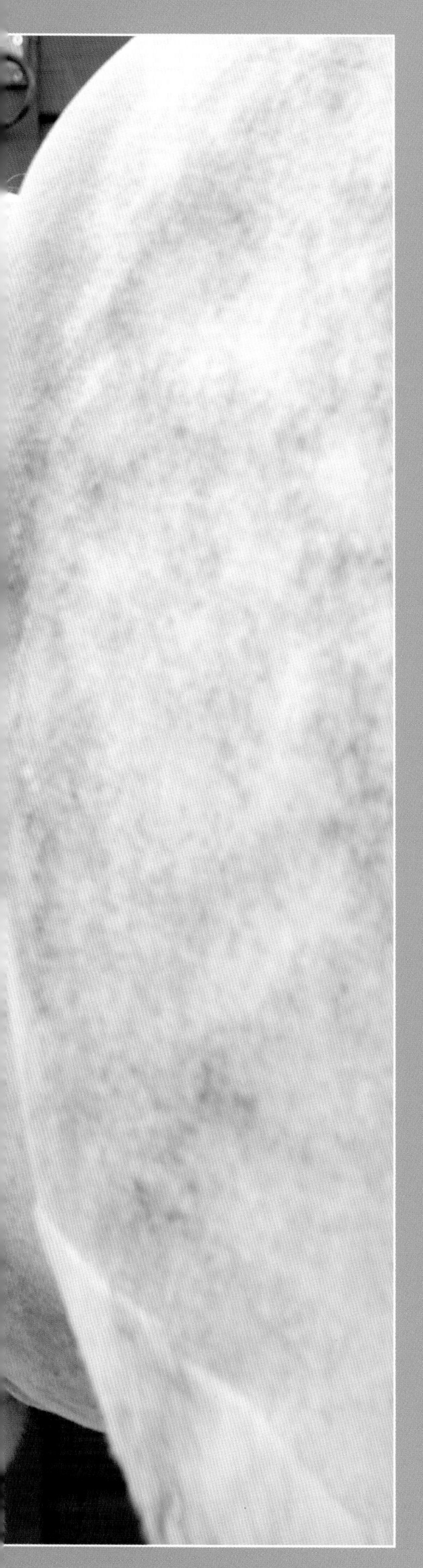

PART 3

CARING FOR A HORSE

Looking after or helping to look after a horse is not only rewarding in its own right, but will help you learn to get to know and understand him. It's also a huge responsibility, so until you become confident and gain enough experience, you need expert supervision and back-up instruction.

Although you can learn a lot from books like this one, you need hands-on experience and expert guidance; there are several ways to get this. Many riding schools run courses for adults and children where you can gain hands-on experience. Throughout the world, you can find schools and stables affiliated to the American Association of Riding Schools, British Horse Society, or Association of British Riding Schools.

Riding and pony clubs may also offer courses or be a good starting point for finding them. The British-based Pony Club is a great organization for young riders with or without their own ponies and has branches throughout the world, offering everything from lessons and rallies to shows. Friends with horses or ponies may offer to let you help look after them. This can also be a way of gaining practical experience, as long as the owner is knowledgeable and the horse is suitable for an inexperienced person to handle. The advantage of starting with the help of a recognized school or club is that the instructors should be qualified, know how to teach, and be aware of safety precautions.

This isn't meant to imply that horses are dangerous—contrary to the old saying, it isn't true that one end kicks, the other end bites, and what happens in the middle is just as bad! But it is all too easy to get into a dangerous situation when you're inexperienced, so minimize as many risks as possible.

Where do I keep him?

Although everyone dreams of keeping a horse at home, most people have to rent grazing and stabling on a professionally run yard or barn. Even if you have the facilities at home, it's a good idea to start off on a good professional yard so you have help and advice available. It also gives you the chance to meet other owners and establish a good, friendly horsey network.

Keeping a horse this way is often known as keeping him at livery or boarding him and the yards that provide such services are called livery or boarding yards. There are several types of livery. DIY is, as the name suggests, where you do everything yourself. Part-DIY or assisted livery means that you do some of the work and the yard staff perform other tasks and full livery is where the yard staff care for your horse in every way so that if you want to, all you have to do is ride him.

Grass livery means that the horse lives out all year round and doesn't have access to a stable. It can work well as long as the fields are well looked after, has good shelter, and you are guaranteed temporary access to a stable if your horse or pony is ill or injured. Although the land owner will usually look after jobs such as applying herbicides and fertilizing when necessary, you will probably have to keep the field clean by removing droppings. Grass livery is usually offered on a DIY basis, though some help may be available.

For most new owners, part-DIY is practical and affordable. Most of us have work and/or family commitments and this system means that you should be able to guarantee that your horse gets all he needs, when he needs it. It also means that someone more experienced is checking him regularly.

When you're looking for a yard, find one where you think you as well as your horse will be happy. For instance, someone who is becoming a horse owner for the first time might feel out of place on a yard where the other owners are all competing at high level and be much happier on one where the clients have a wide range of interests, from hacking (trail riding) to competing at local shows.

Don't be blinded by appearances. While the environment must be safe—you don't want to see gates tied together with bits of rope or ramshackle fencing—finishing touches such as hanging baskets of flowers, no matter how nice they look, don't add anything to your horse's welfare. Your priority should be a basic, well-maintained yard that is kept in a safe and tidy condition, with fields to match.

As you gain more experience, you might prefer to take full responsibility for your horse's care. This may mean either devoting more time to traveling to the yard and doing the work, or teaming up with another trustworthy owner or owners to establish a help rotation.

A Good Living

Whether you are keeping a horse on a professional yard or at home, you need to be able to recognize what's needed. If you are finding a yard on which to keep him, you also need to be able to judge whether standards are acceptable or not.

As explained in the previous section, a horse needs to be kept in an environment that takes his mental and physical needs into account. Basically, this comes down to well-managed grazing (which, depending on his type and susceptibilities, may need to be restricted sometimes) with safe fencing and shelter and suitable companions.

Shelter in the field is important all year round. He needs it for protection against wind and rain in bad weather, and to get away from flies and other biting insects in warmer times. If you are lucky enough to have a field with trees and

thick hedging, this may provide enough natural shelter; if not, a purpose-built shelter should be provided.

The field must also have a safe, constant supply of clean water, in a trough or other suitable container. The container must be free from sharp edges on which the horse could injure himself and should be checked and cleaned out regularly to remove debris and algae.

A stable will also, of course, provide shelter and more information on buildings is given later in this section.

A Growing Concern

Your horse's field is a valuable resource, as it provides food, freedom, and the chance to socialize with his companions. In theory, the larger the field, the better it is, as its occupants will have more space to roam. In practice, grazing land is limited in most parts of the world.

However, you must have enough space to provide enough grazing and allow horses sufficient individual space. The recommended minimum is one acre (2.47 hectares) per horse with one acre to spare, but this is a general "rule of thumb." If the land is well managed and you are keeping small ponies, you may manage with less.

So what does "well managed" mean? A field doesn't have to look like a lawn, but it must provide adequate grazing—which can be supplemented by feeding extra forage outdoors if necessary—and be free from hazards. Potential dangers range from poisonous plants to rabbit holes (which may cause injury if a horse puts a foot in one), rubbish which has been thrown in from outside, and damaged fencing.

Unfortunately, horses don't make it easy to keep grazing in good condition. They are picky eaters and prefer short grass to long. This means that unless the grass is kept at an overall even length by topping (cutting) it at peak growth times, you will end up with a mixture of long, unpalatable patches and others which are overgrazed.

Horses also designate certain parts of a field as toilet areas and won't eat the grass there unless there is no alternative. This is one reason why it's important to remove droppings regularly: ideally, every day. Another even more important reason to "poo pick" a field is that it reduces the worm burden.

Poisonous plants can literally be a killer. The most common is ragwort. When eaten in the field, or in hay which has been made from a ragwort-infested field, it causes irreparable liver damage. Once this happens, a horse must be euthanized to save him from a slow, painful death.

The answer is not to put him at risk in the first place. Learn to recognize ragwort in its growing stages and dig up all plants, then carry them away and burn them. Always wear gloves when uprooting or carrying ragwort, as the toxins in it are poisonous to humans as well as horses.

Ragwort is biennial and forms flat rosettes in the first stage of growth and yellow flowers in the second. Its seeds can lie dormant in the ground for up to twenty years. Other dangerous plants include bracken, foxgloves, members of the nightshade family, and yew. Acorns can also be poisonous, so if there are any oak trees growing in your field they should be fenced off in autumn and winter or horses grazed elsewhere.

There are several types of fencing that are suitable for horses, from traditional to high tech. Thick hedging is ideal and also provides some shelter. Post and rails is a traditional favorite and other safe materials include stud fencing—a type of wire mesh designed to be safe for horses, usually

Far left: and left Ragwort forms flat rosettes in the first stage of growth and yellow flowers in the second. It is lethal to horses and must be dug up, carried away and burnt.

with a wooden rail along the top and not to be confused with sheep fencing—and electric fencing. Electric fencing can be permanent or temporary. The temporary kind is easily moved and is useful for splitting large areas. It is not safe to use it as a boundary fence.

High tech fencing made from special "plastic" is durable, attractive, and virtually maintenance free, but expensive to install. Occasionally, you may see high tensile steel wire used to fence horses' fields, but this is not recommended, as if it breaks, it recoils at high speed and can cause serious injury. Plain wire is not recommended, as there is also a potential for injury, and barbed wire should never be used with horses, because it may be the cause of serious injury.

As anyone who owns horses will tell you, they sometimes seem like accidents on four legs waiting to happen. Minimize risks by making sure that you or someone on your yard checks your horse's field every day. As well as keeping an eye out for damaged fencing and poisonous plants, you can fill in any rabbit holes and pick up any rubbish that has been thrown over the fence by thoughtless passersby.

Clockwise from left: Post and rail; stud fencing; a combination of hedging and post and rail and electric fencing. These are all useful and safe options for keeping fields safe and secure.

Shelters and stables

You don't necessarily need to feel sorry for your horse when the weather is cold, as horses are happy in much lower temperatures than we would find comfortable—though there are times when he may need blankets. However, they do hate a combination of wind and rain and are also bothered by flies and biting insects in milder weather.

A field shelter will give your horse protection from both extremes and can range from the luxurious to the functional. Most commercial designs are similar to stables, with roofs and front openings, but a simple cross shelter comprising four barriers set in the shape of a cross is cheap and effective. With standard designs incorporating front openings, make sure that the openings are wide enough to prevent a dominant horse from trapping others inside or out.

Stables can be individual buildings, sometimes called loose boxes, or internal

A simple cross shelter is cheap and effective.

External stables, or loose boxes.

stables set within a large building, an arrangement which is often called an American barn. Loose boxes get their name because the horse is free to move around, in contrast to old-fashioned stalls, where horses were tied up and could stand up and lie down, but not move around.

Any building which houses horses should provide a healthy environment, with good ventilation and airflow and minimal dust and mold spores. Bedding and management play an important role in reducing dust and mold, but building design is also a factor. A specialist construction company will make sure that a building is constructed or adapted to allow good ventilation without drafts; if possible, have openings in the back as well as the front of a stable to facilitate this and give your horse a choice of views.

If you only have one stable door, split into two, never close the top door unless under veterinary instruction to do so. Your horse will not only be unhappy because he can't see out, he will be breathing in dust and also ammonia fumes from wet and soiled bedding. If you need to keep him warmer, use a different blanket or add an extra one.

Good drainage is another factor to consider. Even the smallest pony can produce an awful lot of urine in a few hours! If your stable allows it to drain away into a suitable outlet, you will have a healthier environment and will save on bedding costs.

Compatible horses with no health issues are often happier if there is a grille in the top half of the adjoining wall. This allows them to see and "talk" to each other. If this isn't possible, many are more contented if you put in a stable mirror made from specially toughened material. The theory is that seeing their own reflection gives the illusion of company.

Internal stables set within an American barn.

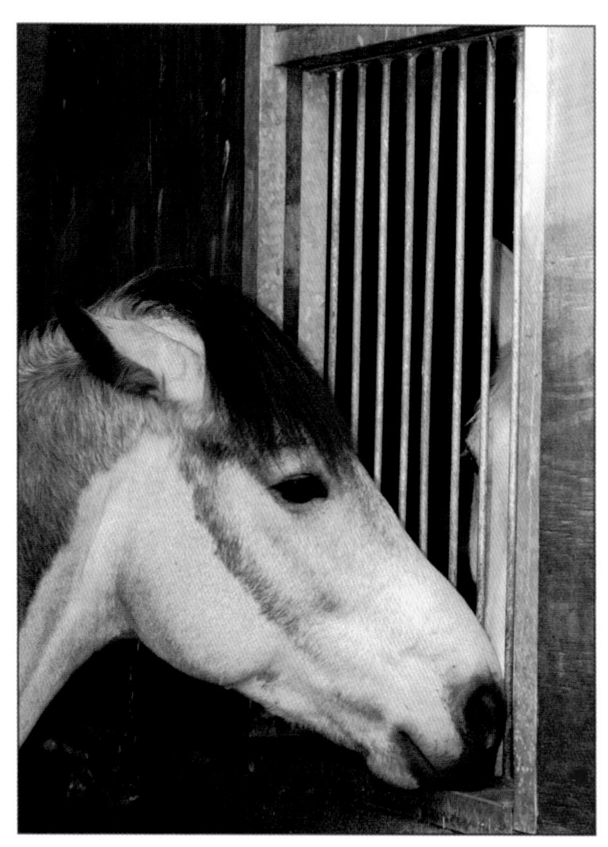

A dividing grille allows compatible neighbors to see and talk to each other.

Bedding & mucking out

Your choice of bedding and the way you muck out are important factors in creating a healthy environment. At one time, horses were nearly always bedded on wheat straw and some people still use this. However, it can be dusty and it also contains a woody, indigestible substance called lignin—so if your horse likes to spend all night munching his way through his bed, he may be at greater risk of colic. This is a serious digestive problem that in the worst case scenario can prove fatal.

There is a huge range of bedding materials, some of which are described below, together with their pros and cons. You also need to decide on your system of mucking out. This can be full, semi deep litter, or deep litter.

A full muck out means that droppings and any wet bedding or soiled bedding is removed daily and replaced with fresh. Semi-deep litter means that droppings and obvious wet patches are removed daily but the bottom layer remains undisturbed, fresh bedding being added on top as needed. Once or twice a week, the stable should be mucked out completely and fresh bedding added.

With the deep litter system, droppings are removed daily but the rest of the bed is not disturbed, fresh being added on top as needed. After several weeks or months, the whole bed is taken out and a fresh one started.

Wheat straw

A full muck out is the healthiest option but is most expensive in terms of bedding costs. Semi-deep litter can be practical and still healthy, depending on the type of bedding used. Deep litter is usually unhealthy because of the build-up of ammonia, which will affect the horse's respiratory system, and because standing on a build-up of dirty bedding can lead to an infection of the foot called thrush.

The exception to this is if you use a relatively new idea, an enzyme product which is applied to the bed at regular intervals and which "eats" harmful bacteria. This can only be used with wood shavings or other bedding materials, according to the manufacturers' recommendations, and means a bed can be left down for several months without compromising the horse's health. If you try this, you will find that when the bed is finally dug out—which takes a lot of hard work or the use of a skid steer—there is no unpleasant smell.

Bedding Materials

Wheat straw is relatively cheap and usually easy to obtain. Because straw is now combined into shorter lengths than was standard when everyone used this bedding, it is no longer as hard wearing. Straw invariably contains dust and takes up a lot of storage space as well as creating large muck heaps; it must be stored in a way that keeps it clean and dry.

Miscanthus (elephant grass) is now being grown both for animal bedding and as a fuel product. It is more absorbent than ordinary straw but not as absorbent as shavings and other materials. As much dust as possible should be extracted before it is packaged.

Wood shavings are often a by product of the building industry, so price depends on the healthy or otherwise state of the latter. Bedding manufacturers screen shavings for foreign objects such as nails and extract as much dust as possible. Shavings are absorbent and easy to handle and store.

Hemp

Wood shavings

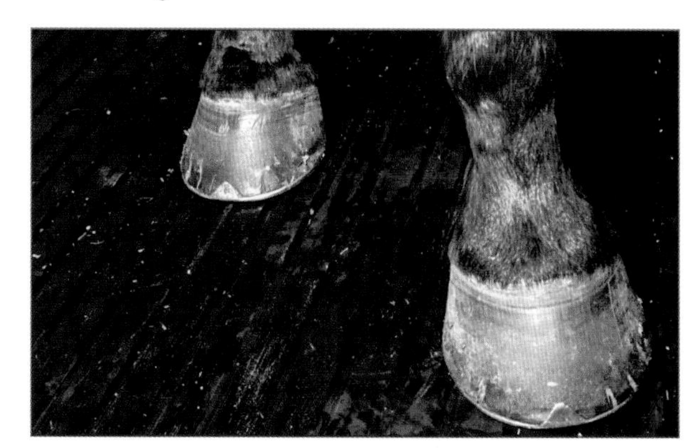
Rubber matting

Hemp is absorbent but, like any organic crop, some bales may contain more dust or spores than others. As always, not recommended if a horse shows sign of eating it—some bedding products swell when wet and if this happens as part of the digestive process, could cause problems.

Shredded or chopped paper or cardboard is free from dust and spores, but heavy to lift when wet and not always easy to dispose.

There are several wood or straw pellet products which are designed to be sprinkled with water, which makes them expand. Because the bagged products are relatively small, they cut down on storage, but they may be relatively expensive.

Rubber matting has become universally popular. It is best used with bedding such as shavings on top, as some horses don't like to urinate on matting alone because of the splash factor—and who can blame them? However, you can use a smaller amount of bedding than usual, because the matting removes the chill from a concrete floor and also has a cushioning effect.

Muck In and Muck Out!

Although some people find the idea of mucking out a stable unappealing, it isn't unpleasant and the "pong factor" is negligible. With the right tools and a bit of practice, you'll soon have it down to a few minutes' work.

You'll need a wheelbarrow that's easy to maneuver, a broom with stiff bristles, a shovel—one with a strong plastic blade is lighter to handle and won't damage rubber matting—and a fork suitable for the bedding. If you are using straw, a four-pronged fork is best for lifting

bedding and a two-pronged pitchfork makes it easier to scatter new bales.

For shavings, you will need a shavings fork, where the prongs are close together. Manufacturers of other types of bedding sometimes sell forks designed for particular ease of use with it. Add a strong pair of rubber gloves and you are ready to go.

If you are starting a straw bed from scratch and don't have rubber matting on the concrete base, it should be thick enough that when you stick a fork into it, the prongs do not touch the floor. With shavings and other materials, aim for a similar depth, so the horse doesn't scrape through to bare concrete when he gets up. Traditionally, there should be deep banks at the sides of the stable, as this is often thought to help prevent the horse getting cast. This is when he lies down near a wall or rolls and becomes jammed against it when he tries to get up. However, it's debatable whether banks are much help in this respect, though they will help prevent drafts.

Whenever possible, take the horse out of the stable before mucking out and either tie him up outside or turn him out in the field. On some yards, where there is a large number of horses and stables are mucked out at the same time, it is standard practice to tie up horses in the stable and muck out round them. However, while it removes the risk associated with horses standing close together and nipping at each other in play, or perhaps kicking out at a neighbor, it means they are exposed to dust and spores thrown up in the air.

Some horses are cleaner in their habits than others; if yours is one who always uses the same corner, you're lucky. Start by removing droppings and any obvious wet patches. If you are making a full muck out once every week or two weeks, that should be sufficient—all you need to do is top up with fresh bedding if necessary.

Start a full muck out by taking out droppings, then sift through the bedding removing wet and soiled material. Work through the sides of the bed, sifting and loosening it and piling it up, then pile the clean bedding from the center on top. Sweep the floor and allow it to dry if time allows before putting the bedding down again.

Put the clean, used bedding in the center and if necessary, add fresh to the sides. The sides of the stable should stay dry, but every couple of days, move the bedding aside and sweep that area too.

With practice, you'll work out your own system, but the one above is a good base. Some people prefer to pick up droppings by hand from shavings and similar beds—which is where the rubber gloves come in!

Feeding

Horses are often described as stomachs on four legs, because given the chance, they spend most of their time grazing. This isn't because they are greedy, but because of the way their digestive system has evolved. In the wild, they travel many miles each day eating grass and vegetation that, compared to the fields in which we keep domesticated horses, is poor quality.

Although it's impossible to keep horses exactly as nature intended, we should try and mimic a natural lifestyle. Horses have a physical and psychological need to spend a lot of time eating forage (grass, hay, and/or haylage) rather than eating a meal, not eating for several hours, then eating another meal.

A horse who is confined in a stable with nothing to eat for long periods will be unhappy and will often suffer both physical and mental problems even if he isn't underweight. He may get colic, stomach ulcers, or develop unwanted behavioral patterns. For this reason, the stabled horse must have access to forage.

Getting advice

Scientific advances mean we know much more about equine nutrition than we did even ten years ago. Although it may sometimes seem that you face an information overload, it's still possible to keep things simple by basing every horse's diet on forage and, if necessary, getting advice from an equine vet or nutritionist about whether you need to add extra feed.

Most feed companies employ nutritionists who will give free advice and help you format a suitable diet for your horse. Although recommendations will be based on their own products, the principles adhered to should be the same whoever you ask. There are also independent nutritionists, who will charge for their advice.

At the same time, you need to make sure that your horse or pony doesn't become overweight. This has detrimental effects on his health just as it does with people and welfare organizations that say obesity is becoming a big problem in more ways than one.

Weighty Matters

The scientific way of assessing a horse's condition is through a system called condition scoring. There are several scoring systems, but the easiest to use ranges from one (poor condition) to nine (obese.) Aim for a perfect five, which means that the ribs can't be seen but can be felt and there are no pads of fat on the crest of the neck and no obvious gulley along the back.

It's useful to know and assess your horse's weight regularly, because otherwise you might not spot gradual changes in weight. The only accurate way to do this is on a large weighing scale—you'll find these at many veterinary practices—and if you get the chance, use one.

You can also use a weightape, which isn't strictly accurate but, if positioned correctly each time, will show you if your horse is gaining or losing weight. These are available through feed companies and in many tack and feed shops.

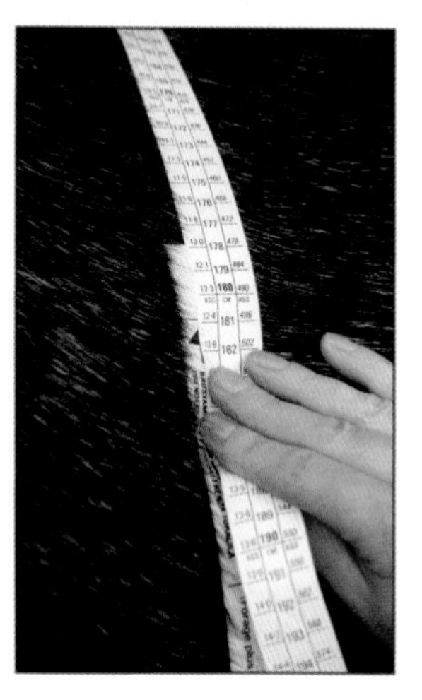

Types of Feed

Forage

As already explained, most or all of your horse's diet should comprise forage. The three basic types of forage are grass, hay, and haylage. Hay is grass which has been cut, dried, and then baled. Haylage is grass which has been cut and either sealed in plastic wrap or bagged while it retains some of its moisture.

Although haylage usually has a higher feed value than hay, it contains more water and less fiber. Both hay and haylage can also be made from alfalfa. This has a high protein value so is not suitable for all horses; for instance, it is too rich for cobs and native ponies. Choose haylage with a feed value suitable for your horse and his workload and type and if you want to slow down your horse's eating rate, feed it from a net with a small mesh or mix it with soaked hay.

Grass usually has higher nutritional value in spring and summer, but climate change means feed values stay high while there is a combination of warmth and moisture. Hay should be steamed or soaked in clean water to minimize risks from dust and mold spores. Haylage usually has a higher feed value and does not need soaking, but has a lower fiber content. Get expert advice on choosing the best forage for your horse or pony.

Hay and haylage can be fed from the ground or in a haynet or specially designed container. Feeding from the ground is better for your horse's dental health, as his teeth will meet at the same angle as when grazing, but can be wasteful. Most people feed from a haynet; this should be tied as shown, high enough to prevent the horse getting a foot caught in it when it is empty and hangs lower.

Other types of feed and products which you may see in feed stores include the following:

Pellets/cubes

Concentrates

When a horse needs more energy than can be supplied by forage, this is usually provided by concentrates, also known as hard feed. These are formulated from cereals and other ingredients in the form of pellets or sweet mix and contain all the major nutrients.

Coarse mix

Again, get expert advice on whether your horse needs concentrate feeds and if so, what type. A feed formulated for a competition horse or racehorse will not be suitable for one doing less demanding work.

Because they take out the guesswork, commercial feeds—which always have a guaranteed nutritional content—are a safer bet for most owners than making up their own feeds from straight cereals and other feedstuffs.

Chaff/chop

Sugar beet

Chaff/Chop

This is the name for chopped forages such as hay, oat straw, and alfalfa. They can be mixed with concentrates or fed alone as a carrier for feed additives such as vitamin and mineral supplements.

Sugar Beet

Sugar beet cubes or shreds comes from the sugar beet, a root vegetable, and are a byproduct of sugar production. They are a good source of fiber and must be soaked according to the manufacturer's instructions, as they swell when wet.

Feed Additives

Often known colloquially as supplements, these range from broad spectrum vitamin and mineral supplements to products designed to maintain healthy hooves, joints, the respiratory system, or whatever. There are hundreds on the market and, with the exception of salt and a broad

Add salt to your horse's feed or provide a salt lick.

spectrum vitamin and mineral supplement, they should not be fed without taking expert advice. If you feed the recommended amount of concentrates, your horse will get all his vitamin and mineral needs. The problem is that the recommended amount may be too much if you want to make sure your horse doesn't put on too much weight and you need to feed less. In this case, feed a broad spectrum vitamin and mineral supplement at half the recommended level.

If your horse is on a purely forage diet, feed it as the manufacturers suggest, mixing it with a small amount of chaff. To make sure he has a sufficient supply of salt, either give him a salt lick or add a tablespoon of salt to his feed, twice a day.

Treats

If you need to tempt a horse to eat or give one who is on a restricted diet something low calorie to munch while his neighbors enjoy their feeds, sliced carrots, apples or turnips are ideal as they are high in water. Why waste money on expensive, commercially produced treats designed to appeal to you rather than your horse?

How Much?

A lot of people are confused about how much to feed their horses. This isn't as complicated as it might seem as long as you keep the basic principles in mind, have a reasonably accurate idea of your horse's weight, and use your eyes and your common sense.

Let's use a 15.2hh all-round family horse or pony who weighs 1,100 pounds (500kg) as an example. He probably lives out in the daytime and is stabled at night and works six days a week, doing a mixture of hacking, schooling, and the occasional weekend competition. This counts as light work and most or all of his needs will probably be satisfied

Sliced carrots are a tempting treat.

by forage—for the purpose of calculation, let's say he works harder in summer and needs a little extra fuel.

This may mean that his diet will comprise 90 percent forage and 10 percent hard feed.

If he is in correct condition, he will need 2 to 2.5 percent of his weight to maintain it, giving a daily total of 27 pounds (12.5kg). This gives us a daily requirement of about 25 pounds (11.25kg) forage and 2 pounds (1.25kg) hard feed. Even if he only needs a small amount of hard feed, be fair at feed times—if other horses on the yard get two or more feeds a day and your horse is stabled with them at feed time, split his feed into the same amount of meals and if necessary, bulk it out with sliced apples or carrots.

Grass will provide between half and all his forage needs, depending on the time he spends grazing. When he is stabled, he needs hay—weighed before rather than after soaking—or haylage. This will help to keep him happy by allowing him to mimic grazing behavior.

When you're working out what to feed your horse, remember that he is a horse and not a human! Some owners worry that their horses will become bored by eating the same things, but this doesn't happen. You may occasionally find a horse who prefers coarse mix (sweet feed) to pellets, or—in the case of a nervous, "stressy" animal—is reluctant to eat unless in quiet surroundings—but you'll never find one who is bored with grass.

The advice in this book can only be general. If you are a new horse owner or feel confused about feeding, talk to someone knowledgeable, preferably an equine nutritionist who keeps and rides horses and can mix science with knowledge gained from practical horse care.

Golden Guidelines

Correct feeding is an art and a science, but above all, it is based on sound principles and common sense. These principles apply whether you are feeding a family pony or an Olympic competition horse. As with all aspects of caring for a horse, it's a big responsibility, but try not to get confused by the hundreds of different products in the market place and keep the following guidelines in mind.

Every horse's diet should be based on forage. In many cases, this—plus a broad spectrum vitamin and mineral supplement to fill seasonal or regional discrepancies, added to a handful of chaff—will be all that he needs to eat.

Clean, fresh water must be available at all times. Both water buckets and feed containers must be cleaned out daily.

If he needs extra fuel because he is working hard, you may need to supply this in the form of compound feed from a specialist horse feed manufacturer. Get expert advice on what and how much to feed according to your horse's condition, type, and workload. Good feed companies have their own nutritionists who will help you work out a suitable regime; they will obviously recommend their own products rather than those from different manufacturers, but should adhere to the principles here.

Clean buckets and containers daily.

Only buy the best quality feed and forage and store it so it is clean and dry. Keep all storage containers and areas clean and discourage vermin. Don't use plastic containers such as dustbins as feed containers, as rats can chew through them.

Assess your horse's condition regularly and aim to keep him so he is neither too thin nor too fat.

If you change from one brand of feed or forage to another, do so gradually over several days, adding a small amount of the new and subtracting the same amount of the old. This should avoid digestive upsets.

Feed by weight, not by volume. A scoop of one feed will not necessarily weigh the same as another.

Don't feed over 4 pounds (2kg) hard feed in one meal.

Make sure that you follow a correct worming regime and that your horse's mouth and teeth are kept in good condition. A heavy worm burden or sharp edges on his teeth can cause many health problems and also mean he will not chew and/or digest forage and feed properly.

Grooming and bathing

A routine of daily grooming is important, not just to keep your horse looking smart but to enable you to check for the first signs of physical problems such as small cuts and nicks that may otherwise be missed, or heat or swelling in a limb. This hands-on time with your horse also allows you to get to know him and to recognize when he is perhaps feeling a little grumpy or off-color. Spotting the first warning signs are an essential skill for any horse owner and often mean that small problems can be prevented from developing into large ones.

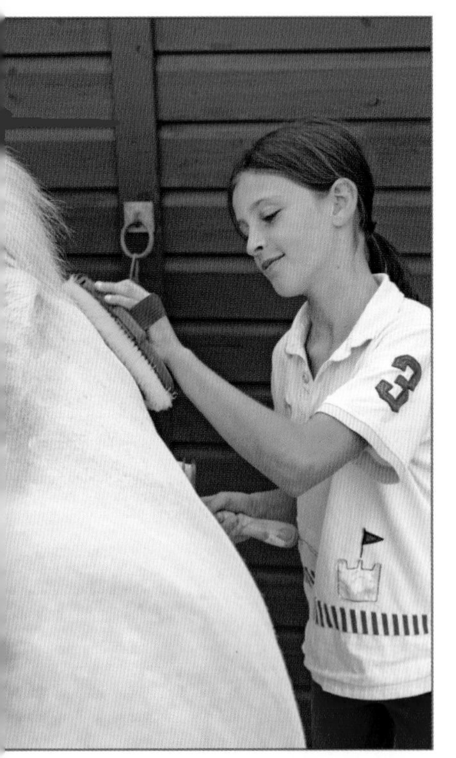

Most horses enjoy being groomed as long as it is done considerately, though some thin-skinned animals are ticklish and need to be treated with extra care; not only do you need to avoid causing irritation or discomfort, you don't want to be nipped or kicked. Watch compatible horses in the field and you'll see them grooming each other, using their teeth on other animals' withers, necks, and backs. Behind and below the withers are favorite spots and your horse will probably enjoy you using a grooming tool with rubber prongs in this area.

Although basic grooming techniques apply to all horses and ponies, the equipment you use and the routine you follow should be adapted to his type and lifestyle. As one extreme, a native pony with a thick coat who lives out all year round needs the natural grease—which looks like white dust—left in his coat, as it acts as a waterproofing agent. As another, a horse with a fine or clipped coat who wears a blanket to protect him from the weather and is stabled and blanketed at night can be groomed more thoroughly.

While the thick-coated pony may be perfectly comfortable if you use a brush with stiff bristles to remove dried mud, a clipped horse certainly won't be. Fortunately, there is a wide range of grooming tools that will enable you to have a clean, comfortable horse.

Choosing a Grooming Kit

All professional grooms have favorite items in their grooming kits, but there are basics that every owner should have. You should also make sure that your horse has his own grooming kit and that it is not used on any other horses. This minimizes the risk of passing on skin infections, notably ringworm.

Hoofpick: Used for cleaning out the hooves, removing mud, and stones. By levering it against the horse's shoes, you can check whether or not they are still secure.

Rubber Curry Comb/Plastic Groomer: Great for removing dried mud and loose hair and for lifting grease and dust to the surface of the coat. Some people prefer to use a dandy brush with stiff bristles for this job, but it isn't as comfortable for the horse. This type of dandy brush should not be used on manes and tails, as it damages the hair.

Whisk or Flick Dandy Brush: A brush with long, flexible bristles that is used to flick dust and dirt away from the coat.

Body Brush: Has short bristles that work through the coat to remove dust and massage the skin. It can also be used on manes and tails without breaking the hair.

Cotton Wool or Cotton Pads: These provide a hygienic way of cleaning eyes, nose, mouth, and dock. They are far better than traditional sponges, as they can be thrown away after each use and thus minimize the risk of infection.

Fly Repellent: Helps keep off irritating and biting insects and is available in various formulations, such as sprays or impregnated wipes.

Extras

You'll soon build up your own box of grooming tricks, but here are some that you'll find in the professionals' kits:

Human Hairbrush: The widely spaced bristles allow you to separate the mane and tail hairs without breaking them.

Cactus Cloth: A plain square or mitten made from coarse weave fabric that is useful for removing stains—or stain removing product.

Stable Rubber or Grooming Mitten: Used to give a final wipe over the coat. Mittens are easier to use around the head area.

Face Brush: A small brush with soft bristles, often made from goat hair, for use on the horse's face.

Strapping Pad: Used to build up muscle on the neck, shoulders, and quarters.

Pulling Comb: A metal or nylon comb with short teeth used for thinning and shaping manes and tails.

Shaping Combs: Combs from the dog grooming industry that offer a kinder way of shaping tails.

Sweat Scraper: Removes excess water after bathing or washing down.

Trimming Scissors: Their blades have curved, rounded ends, so are safer.

Plaiting Kit: Containing elastic bands, plaiting thread and needles, and a dressmaker's stitch unpicker for unfastening the plaits.

Everyday Essentials

There are a few jobs that should be done every day, with every horse, regardless of whether or not he is being ridden. One is to pick out his feet, removing dirt and any foreign objects such as stones, and checking that his shoes, if worn, are secure. Making sure that your horse's feet are kept in good condition is an important part of caring for a horse and keeping him sound.

It's a sign of a horse's trust that he will allow you to pick up a foot, because as a "fright and flight" animal, he is vulnerable when he hasn't got all four on the ground. Always give him warning that you are about to ask him to lift a foot. Don't just grab a fetlock and pull, but run your hand down his leg, starting at his shoulder or hindquarters. Most horses who are accustomed to having their feet picked up will oblige as soon as you squeeze just above the fetlock; hold the foot just high enough off the ground to do the job.

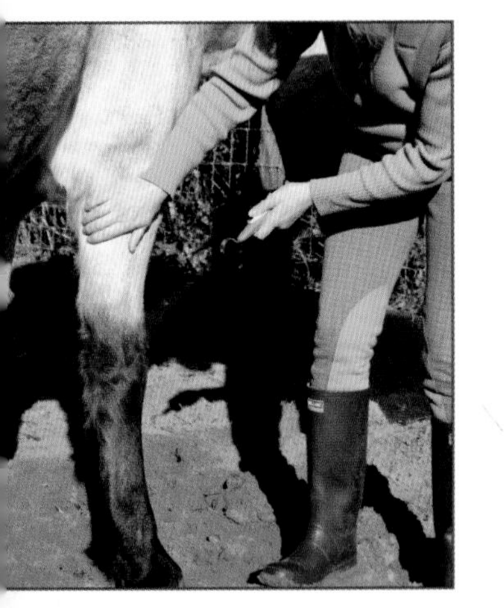

If you have difficulty getting a horse to lift his feet, ask someone experienced to help you. Farriers are the real experts at dealing with ignorant or uncooperative customers.

Use your hoofpick in a direction from heel to toe, to avoid jabbing the sensitive frog (the V-shaped cleft). Don't dig the hoofpick into the frog.

As part of your everyday routine, run your hands over your horse to check for possible problems, as explained earlier. If you are going to ride, make sure that there is no dirt on the areas where tack rests. Dried mud should be brushed off and if he's wet and dirty, you can either wait for him to dry and brush it off or wash it off and blot dry with a towel.

You should also wipe each eye with a separate pad of dampened cotton wool and clean his nostrils and dock the same way. This helps to keep away flies and to keep him comfortable.

Brush-offs

If the horse is stabled part of the time, or needs to be tidied up before riding, you can give him a quick brush over to tidy his coat, mane, and tail. The traditional way of doing this with a blanketed horse is called quartering, because you groom a quarter of him at a time.

Undo his blanket and fold back the front half, then give each side of his front half a quick brush over with a body brush. Any stable stains can be removed by using a cactus cloth or damp sponge or cloth.

Then fold the back half of the blanket towards the front and brush each side of his rear half. Finish by tidying his mane and tail. All grooming techniques are explained below.

Groomed to Perfection

Ideally, all horses except those who live out all the time should be given a thorough groom once a day. The best time to do this is after you have ridden him, as he will still be warm and the pores of his skin will be open. If necessary, put a blanket on him as previously explained so he does not catch a chill.

All owners develop their own grooming routines, but here is one that works well. Working in this order means you're not flicking dirt on to areas you've already groomed, and hopefully, you won't forget anything.

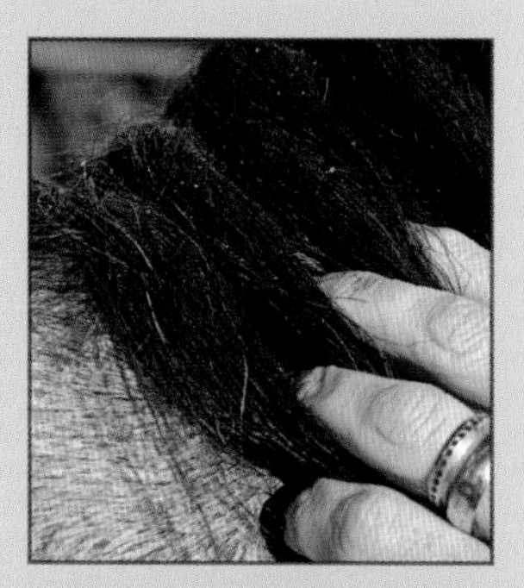

- Run your hands over him to check for any problems and pick out his feet.
- If he has a thick mane, separate it gently with your fingers.

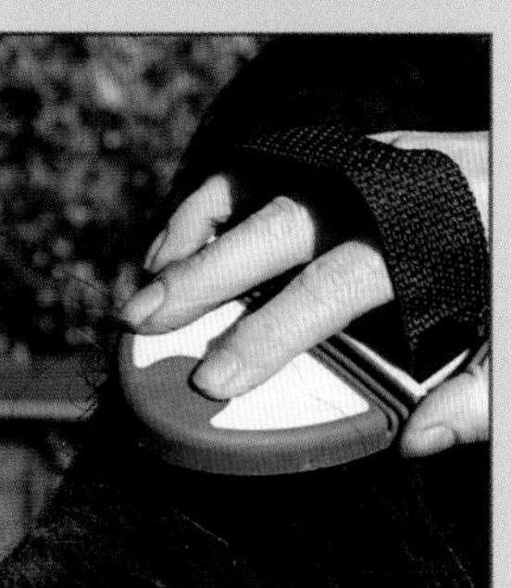

- Brush through both sides with a body brush. Push it over the neck and brush from the roots to the ends, all the way along the neck, then flip the hair back and repeat the same process on the top of the mane.

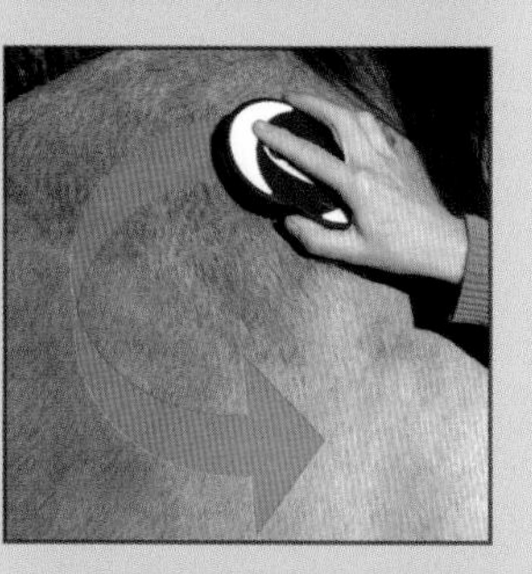

- Starting at the neck, use a rubber curry comb or groomer in a circular motion to remove loose hair and dried mud and bring grease to the surface. If your horse is clipped, you probably won't need to do this.

- Work from top to bottom as you go along the horse but be careful if you use a groomer on the belly area, as he might be too sensitive and show this by kicking.
- Next step is to flick off the dirt you've raised with a whisk dandy brush. Don't apply pressure, just flick away the debris.

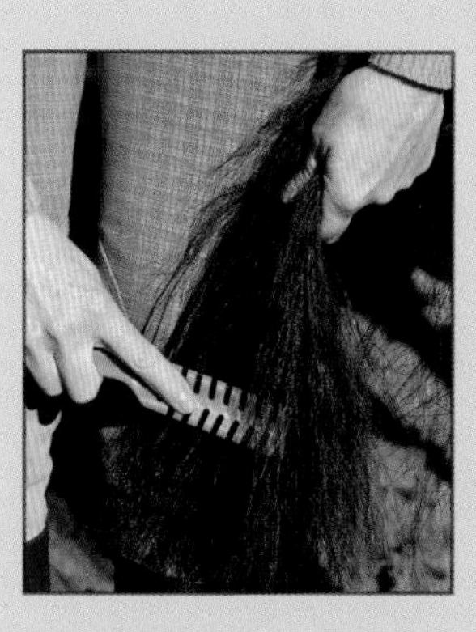

- Separate the tail hairs with your fingers and if necessary, use a body brush or human hairbrush to brush it out. Hold the tail near the ends and brush the bottom section, then gradually move up. By brushing downwards to the ends each time, you won't get tangles at the bottom.
- Now comes the hard work, start brushing at the neck and working back to remove grease and spread natural oils along the hair shafts, which will help give a shiny coat. Use a body brush, held in the hand nearest the horse, and

stand far enough way that you can lean in to the brush strokes. Clean the brush every few strokes by running the bristles over a metal curry comb, doing this in the direction pictured so you don't spread dirt over yourself. Tap the curry comb on the ground to empty the dirt.

- Clean your horse's eyes and nostrils with separate pads, then clean his face gently with either a small body brush or a face brush. Use another pad to clean the dock area.
- Finish by wiping over with a barely damp cloth to remove any dust left on the coat surface.

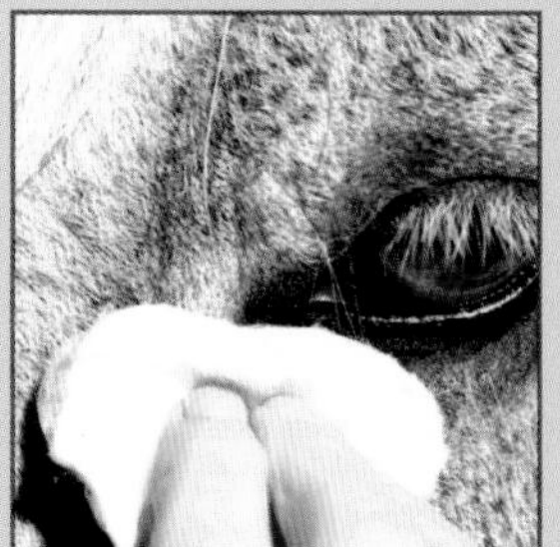

- If weather conditions dictate, apply fly repellent.
- Stand back and admire your handsome horse, who by now is probably a lot cleaner than you!

Personal Hygiene

Geldings and stallions produce a waxy material called smegma in the sheath and penile area. Although there are all sorts of products sold to clean the sheath area, most vets advise leaving it alone unless there is an excess of debris.

In this case, wear disposable gloves and wash the area with lukewarm water. If necessary, use a small amount of mild soap without coloring or perfume, but never use a product containing antiseptic or bacteria, or this may upset the balance of the natural, healthy bacteria and lead to infection.

Mares generally remain clean, but if necessary, bathe the udder area with lukewarm water to remove dirt.

Top Tips

Don't take any chances with ticklish horses. Treat them with respect and wear a hard hat—otherwise, you might end up at the wrong end of a kick when grooming a sensitive area. If you know your horse dislikes having his belly groomed but need to remove mud, use the flat of your hand to dislodge the worst bits.

There are all sorts of cosmetic products available, such as spray-on coat gloss. While it's find to use these for a show, don't use them every day. When you do use them, don't apply them in the area where the saddle and girth rest, as they make the hair slippery.

However, they are very useful to spray on tails and prevent them getting tangled up and can be applied as often as needed.

Hoof oils and dressings can also be applied for competitions and special occasions, but use a product that will not interfere with the natural moisture balance in the hoof. If in doubt, ask your farrier's advice.

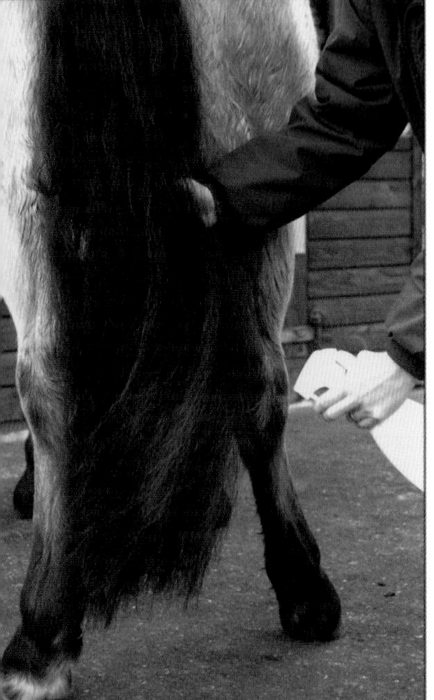

Bath Time

Although bathing is not a substitute for grooming, there will be times when you need to do it, either before a competition or to make the most of a hot day and get rid of dust, dirt, and skin flakes from your horse's skin and coat. Don't bathe a horse on a cold day unless you have the luxury of an indoor wash box and drying lamps, or you risk him catching a chill.

It's best to use shampoo formulated for horses, so as not to irritate the skin. You can also use shampoo made for people, but this can work out more expensive. Don't use washing up liquid, washing powder, or anything else containing detergent, as this can affect the skin and dull the coat. Some horse shampoos are formulated to "enhance" or "brighten" coat colors and the ones for greys seem particularly effective. If you use one of these, do a patch test first just in case your horse is one of the rare animals who shows signs of allergy.

You'll need shampoo; a hose pipe or a helper who will bring a constant supply of buckets of water; a wash brush or sponge for the body area and separate sponges for the face and tail; a plastic scraper to remove excess water and—unless it's hot enough to walk your horse around to dry off—and a thermal blanket which transfers moisture from the coat to the outer layer of its fabric.

Don't take anything for granted when bathing even a quiet, experienced horse. Always stand at his side, never directly behind him, and never kneel down; if necessary, bend or crouch so you can move away sharply.

Horses are sometimes nervous of hose pipes at first. The best way to accustom him to it is to start with a gentle trickle played on his foot, then work up the leg and shoulder until he is confident enough for you to play it on his body. Increase the water flow gradually and don't aim a hose at his ears or face—use a sponge to wet the facial area.

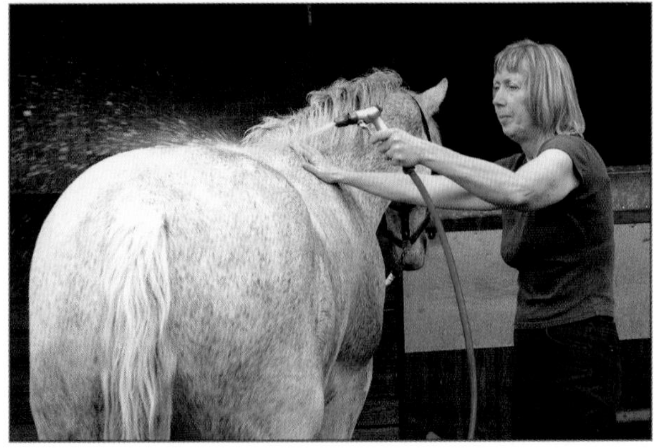

- Starting on the neck and working down and back as when grooming, wet the coat thoroughly, then apply the shampoo, either undiluted or diluted according to the manufacturer's instructions. A wash brush, with bristles surrounding a central sponge, is the most effective applicator for large areas, but when working shampoo into the roots of the mane it's best to use your fingers. Pull the forelock back through the ears to avoid getting shampoo in the horse's eyes.

- Tails can be washed separately, as well as when giving your horse a full bath. Wet the tail and apply shampoo, then rinse clear with a hose or buckets of water. Squeeze out excess moisture, then hold the tail just below the end of the dock and swing the long hairs round to remove more water. Be careful when doing this; if you're not sure whether he's used to it, start by swinging the hairs gently and build up.
- Rinse from the neck back, pushing the water through the coat with your hand. Keep rinsing until the water running away from the coat is clear, with no traces of shampoo. When you've finished, remove as much water as possible with a plastic scraper, being careful not to bang bony areas or press too hard, then towel dry, and if necessary, put on a blanket.

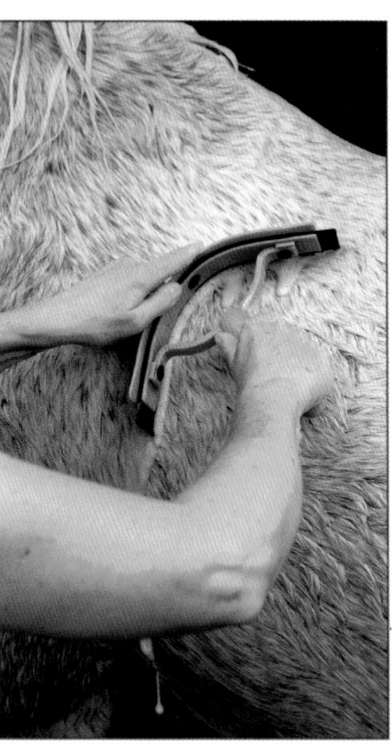

- Finish by sponging your horse's face clean. You'll probably only need to use water in this area, but if you use shampoo, be careful not to get it in his eyes.

Bed Bath

Faced with a dirty horse and not able to give him a full bath, perhaps because the weather isn't warm enough? If he has a fine or clipped coat, a hot cloth wash is a good substitute.

You'll need pieces of clean towels or face cloths and two buckets of very warm water. Soak a cloth and wring out as much water as possible, then rub it on the coat in a circular motion, following the lie of the hair. Rinse the cloth frequently in the second bucket and change the water as often as needed.

On a cool day, you can use this technique on a blanketed horse by folding the blanket as when quartering. Go over each section with a dry towel before going on to the next.

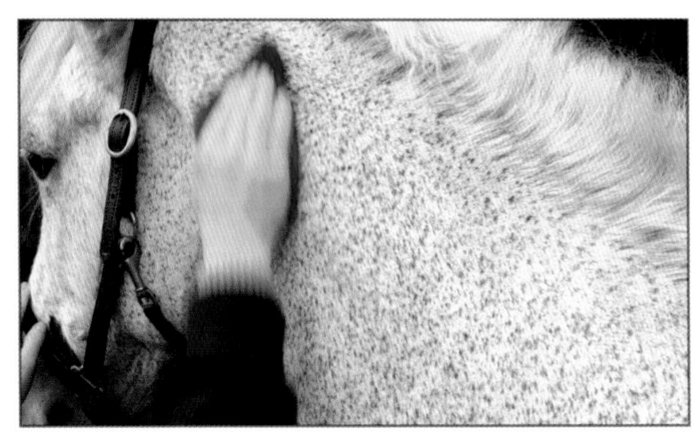

Clipping

Some horses and ponies grow thick winter coats and if doing any more than light work, will sweat excessively. This means you have to be careful to dry them off properly, or they may catch a chill. Others, in particular, Thoroughbreds, have fine winter coats which do not cause any problems when they are on a light to medium work regime.

To keep a horse with a thick coat or one who is in harder work comfortable, it's often necessary to clip him in the autumn and winter months. Clipping also makes a horse look nice, which may be important if he is competing, but this should be a secondary reason for taking off some or all of his coat. It's easy to tidy up a horse just by trimming him up, as explained later.

There are several types of clips and before you opt for a particular one, think about your horse's workload and lifestyle. As soon as you remove any hair, even by giving him a minimal clip, you will need to provide him with blankets to give protection against the weather. The more hair you take off, the more protection he will need; most people prefer to stable horses who have anything other than a minimal clip at night and sometimes, it might be imperative to do so.

The main types of clips, starting with the one which necessitates removing the minimum amount of hair and moving up, are:

- **Bib Clip**: Hair on the chest, underside of the neck, and throat is clipped off.
- **Trace Clip**: Removes hair from the underside of the neck and throat and the lower part of the body. It was derived for driving horses and follow the lines of the harness traces, hence its name.
- **Chaser Clip**: Follows a diagonal line from the ears to the end of the belly, leaving hair on the back and hindquarters.
- **Blanket Clip**: As the name suggests, leaves a blanket of hair over the back.
- **Full or Hunter Clip**: All the hair except for that on the legs is clipped off. In most cases, it's best to leave hair on the area where the saddle sits, called a saddle patch.

Bib clip

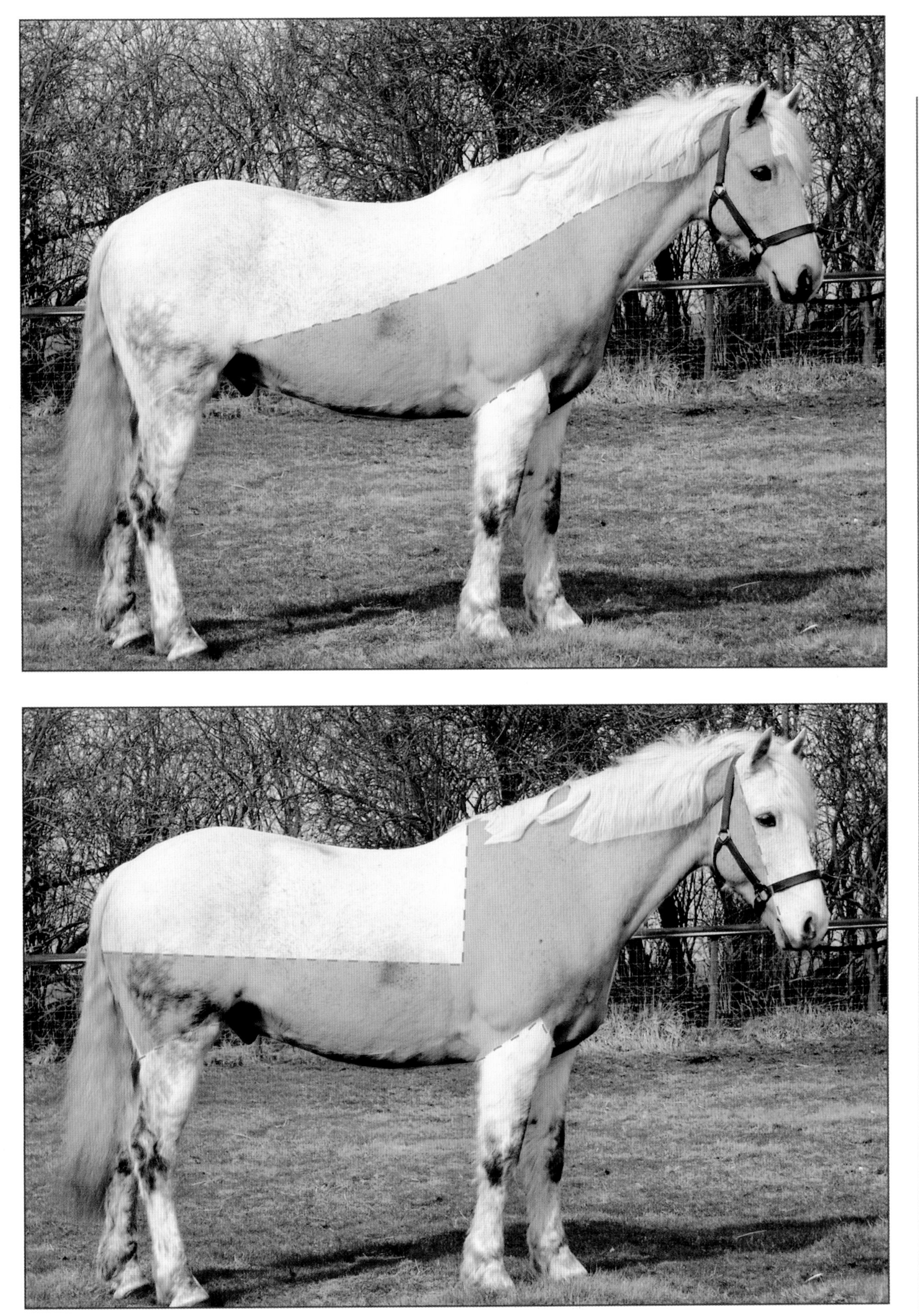

Top: Chaser clip **Above:** Blanket clip

Clips can be adapted to suit. Trace, chaser, and blanket clips can be higher or lower, depending on how much hair you want to leave on. If in doubt, start by removing the minimum amount of hair and if necessary, take off more the next time round.

If you want to be different, you can buy stencils that enable you to clip letters or patterns into the coat on your horse's hindquarters. Popular ones include national emblems, such as Canada's maple leaf and Ireland's shamrock.

You can either buy a set of clippers and learn to clip your horse, or pay a professional to do it. The best way of learning how to clip is to find someone experienced and skilled to teach you the techniques on a quiet horse, but the information here will help you.

Safety is important, so only attempt to clip in safe surroundings. Your working area should be dry and light and the horse must be clean and dry. If this means you need to clip your horse in his stable, pile up all the bedding round the sides so you can sweep out the hair that comes off. Take out water and feed containers or cover up automatic waterers so hair doesn't fall in them.

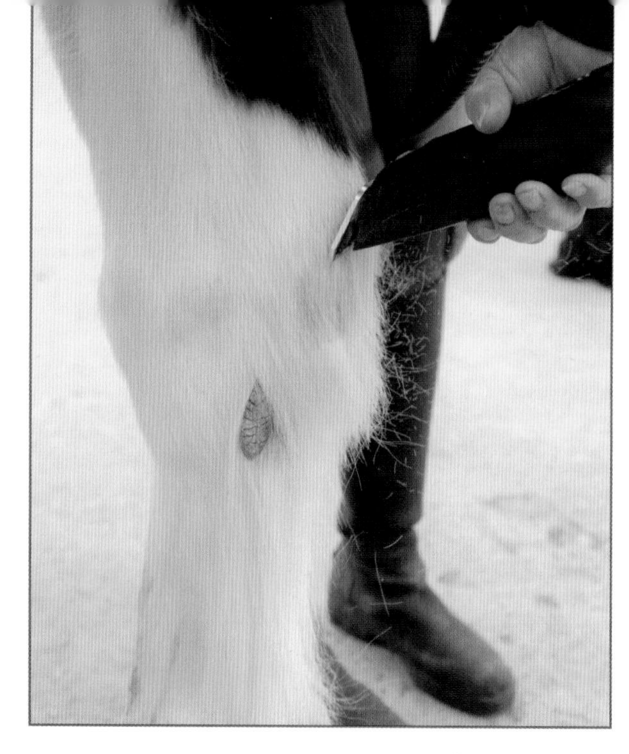

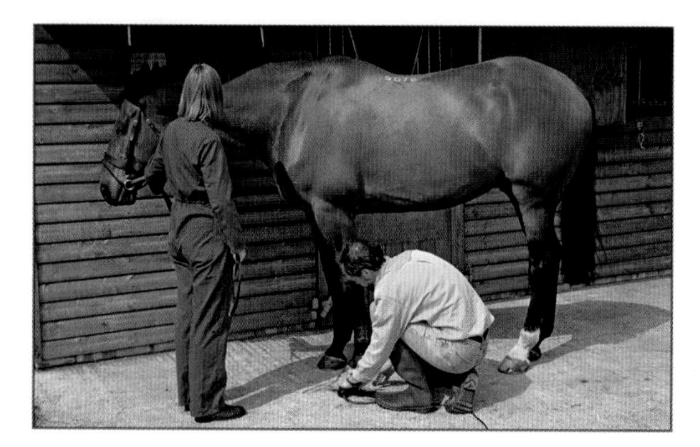

If the weather is dry and still, you can clip a quiet horse outside his stable, but never try and do this if it's windy or starts to rain. Although you can clip at any time of year—some endurance riders may even clip their horses in summer to minimize the risk of them overheating—it's usually best to wait until the winter coat has come through fully. Unless you need to clip for welfare reasons, don't do it once the summer coat starts coming through, as clipping blunts the ends of the hairs and may reduce the coat quality.

Different types of clippers suit different purposes, so get expert advice on choosing the right ones for you and your horse. Battery-operated clippers are usually quieter when running than electric ones, so are useful with inexperienced or nervous horses. However, you need to check that they are powerful enough to cope with the type of clip you have in mind.

The type of clipper blades is also important. They are available in fine, medium, and coarse formats and for most horses, fine or medium ones are best. Coarse blades, which are really intended for cattle, are sometimes useful for clipping horses with very hairy legs.

Blades usually last for six to ten clips before they need sharpening by a specialist; this isn't a DIY job. Never use blunt blades, as they will pull the hair and cause discomfort, which not surprisingly may make a horse wary of being clipped. They are ground to work as a pair, so you can't match half of one pair with another.

Always have a spare pair of sharp blades at hand. If the ones you're using become blunt in use, or you drop and break them, you'll be able to carry on rather than having to explain why you have got a horse who is only half clipped!

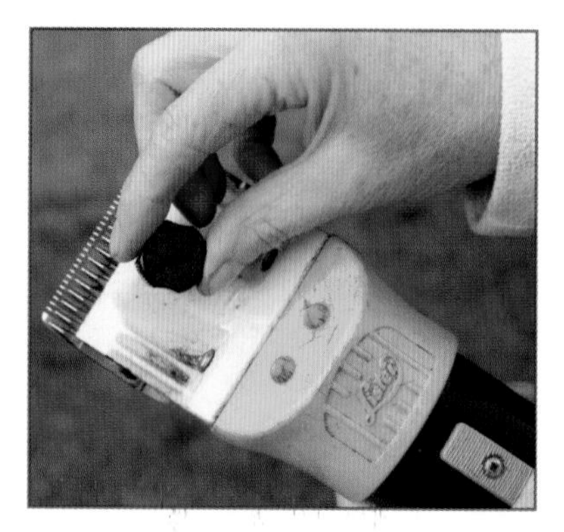

Before you start, read the clipper manufacturer's instructions and make sure the blade tension is correct. Blades should be

lubricated with special clipper oil before you start and at regular intervals to keep them running smoothly. Don't use any other type of oil, as it may irritate the horse's skin or cause the blades to clog up.

Are you safe to clip? If you're not sure how the horse you are going to clip will behave, or know he is likely to be nervous or ticklish, wear a hard hat as protection against possible kicks to the head—a horse who is ticklish in the belly area can easily catch you with a kick as you lean down. If you are using electric clippers, always use a circuit breaker unless your power supply has one built in and be sure to wear rubber-soled boots.

Anyone with long hair should tie it back to remove the risk of it getting caught in the blades. As clipped off horsehair gets everywhere, it's sensible to wear overalls. If they don't have cuffs, put elastic bands over the sleeves to prevent at least some of the hair working underneath and if you don't want to wear a hard hat, a baseball cap or similar cap will keep your horse's hair out of your own.

Some people will happily clip a horse on their own, but it's always easiest and safest if you have a helper. If necessary, a helper can reassure the horse and if you're clipping round his elbow areas, you need someone to hold up his leg to keep the skin stretched as you work.

Many owners give their horses a full clip, regardless of whether or not this is most suitable, because they are worried about keeping the lines of another design straight and level. The easy answer is to mark out clipping guidelines on the horse before you start, using a dampened stick of chalk.

Where a line on one side needs to meet up with its partner on the other side, either over the withers or round the back of the hindquarters, use a piece of string to get a perfect match. You can also use chalk to mark out a saddle patch. You can either draw around a saddle pad or just mark out the contact area under the saddle seat; in either case, make sure you position it correctly.

If you're really clever, you can clip to enhance your horse's conformation. For instance, if he is short in the neck, giving him a blanket clip that is set slightly farther back can give an optical illusion of greater length in front of the saddle.

Step by Step

Get everything ready before you start, so you don't keep your horse waiting around and getting bored or restless. Make sure he is dry and as clean as possible and if you're clipping the hindquarters area, it's a good idea to put a tail bandage on him to keep the hairs at the top of his tail out of the way. Set up and lubricate the clippers and have a blanket ready to put over him.

A good professional groom will always take the long-term view, aiming to keep the horse relaxed and happy. You should do the same, even if it means you can't do everything you want first time around. It's better to give an inexperienced horse a small clip such as a bib clip to start with, then extend it over subsequent clips, than to risk frightening him and setting up an argument for next time round.

No matter how quiet and experienced the horse, talk to him and run the clippers for a short while so he gets used to the noise. Check that cables are out of the way of his and your feet and prepare to start clipping on an area that's usually less sensitive: a shoulder is ideal if this fits in with your proposed clipping pattern. Before you put the blades on the horse, rest your hand on his shoulder and lay

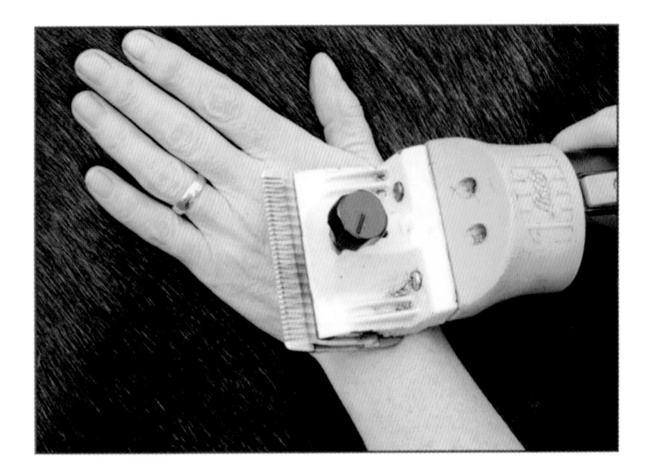

the clippers on top to accustom him to or remind him of the vibration.

Always clip against the lie of the hair, keeping the blades flat against the coat and using an even pressure. You don't need to press hard, but the pressure should be firm. Think about making reasonably long strokes rather than short ones and overlap each stroke to get a smooth finish.

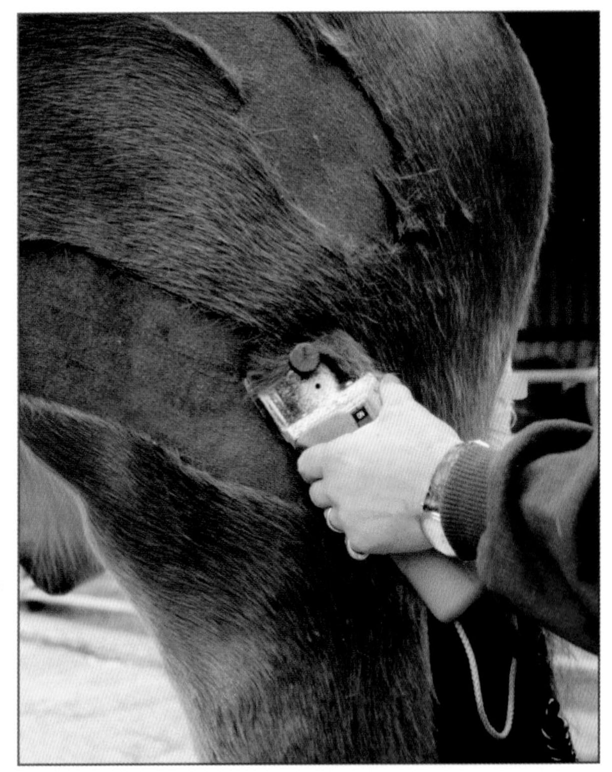

Although most clippers have air vents that blow out cool air over the blades to help prevent hair clogging them up, you need to stop clipping at regular intervals to check that the blades aren't heating up and that the vents are free. There will be times when you need to use a small, soft brush to clean the vents and you will also need to lubricate the blades regularly.

If the running noise of the clippers changes, or they start running more slowly, turn them off immediately, clean off any hair, and lubricate the blades.

As with grooming, everyone develops a preferred clipping routine, but plan it so that you avoid the horse getting cold and minimize any anxiety. If you get to an area and he starts getting worried, it's sometimes better to do a little bit, move on to an easier bit, and then go back to the "trouble spot" when he's relaxed again.

Once the horse is happy with the first few strokes of the clippers, you can follow your chalk lines to mark out the shape of your clip. To get a neat finish, clip up to just below the line, to give a little leeway, then if necessary, hold the blades at an angle to straighten any wobbly bits.

Start clipping at the front of the horse and as you take off more hair, put a blanket over his loins and quarters to keep him warm. Even if it isn't cold, he's going to feel the difference. As you move to the back end, fold the blanket forward as when grooming.

If possible, it's best to leave some hair on your horse's legs to give protection. Follow the line of the muscles for both the front and hind legs. With the front legs, this will give an upside down V and with the hind legs, you'll have a line sloping towards the stifle. If you're not quite sure what would look best, draw chalk lines to give you an idea.

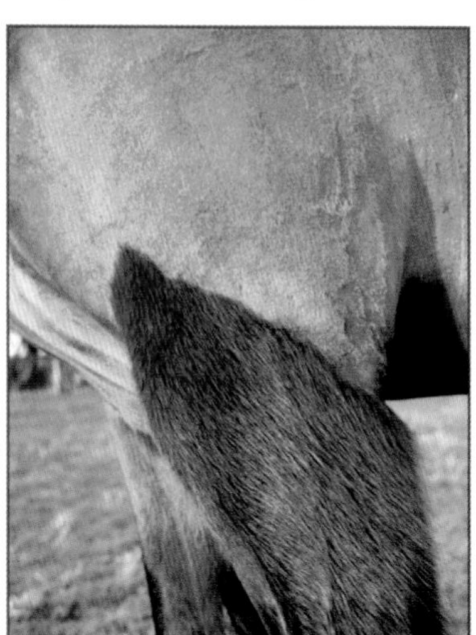

You'll find that in some areas, your horse's coat grows in an irregular way. Most horses have whorls, small patches where the coat radiates from a central point and you'll often get a line on the chest where hair growing from each side meets in a raised line. The hair above and to the side of the stifle joint also has a growth pattern that radiates out from the joint.

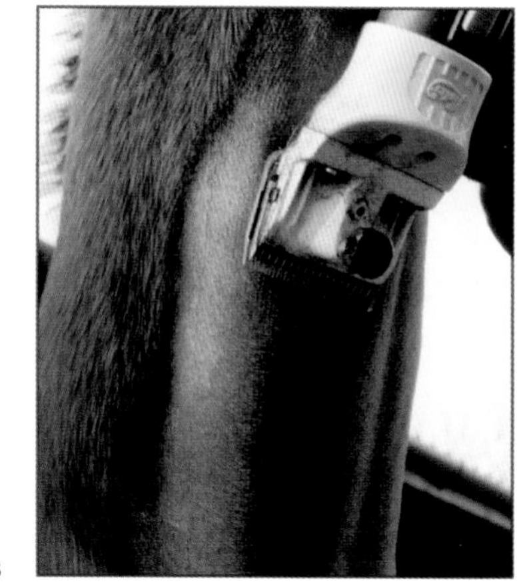

In these places, you need to change the direction of your clip so that you are still working against the lie of the hair. You'll also find it easier in some areas if you gently stretch the skin with your other hand.

A few areas are more fiddly than others. The belly area is notoriously ticklish, so be extra careful when working here. You'll also find that some horses are more ticklish on one side than the other. In these cases, they are often more reactive when their offside (right-hand side) is being worked on. This may be due partly to the fact that we tend to do most things—leading, tacking up, and so on—from the nearside (left-hand side.) Yet another reason to try and handle your horse equally from both sides!

It's essential to have a helper when you're clipping the elbow and girth areas. By picking up each front leg in turn and gently stretching it forward, your helper can avoid the skin creasing into folds and make it easier for you to get a smooth finish.

Be careful not to cut into the mane when clipping a horse's neck. Run the clippers just below the mane line, at a slight angle. If you're not sure how high you can risk clipping, err on the side of caution. If you cut into the hairs at the base of the mane, they will stand out until they grow back long enough to lie flat, which takes some time.

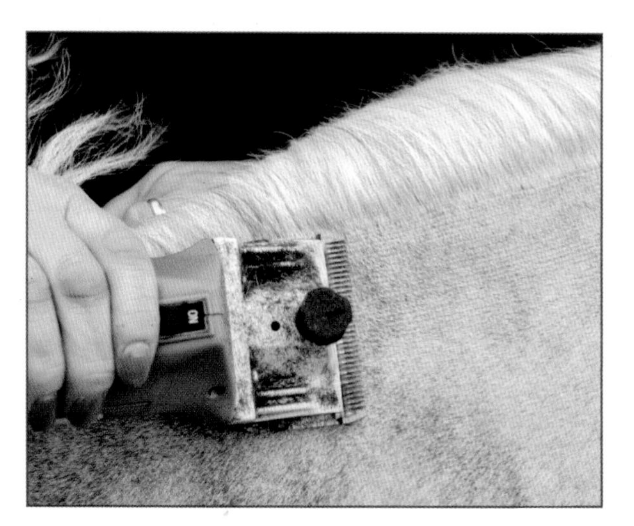

Clipping the head gives the perfect finish to full, blanket or chaser clip, but some horses are wary of clippers near the ears and head—perhaps because the sound and vibration resonates more. However, if you're careful and considerate, most animals learn to accept it.

Experienced grooms may decide to clip the whole head, but because this involves working around the eye area it should only be done with the greatest care and caution and you'll need to use quiet clippers with small blades. Often, a better option is to clip the head area up to a line that goes from the base of the ears to the top of the lips—use a piece of string and draw a chalk guideline. When the horse's bridle is put on, the line will be hidden by the bridle cheekpieces.

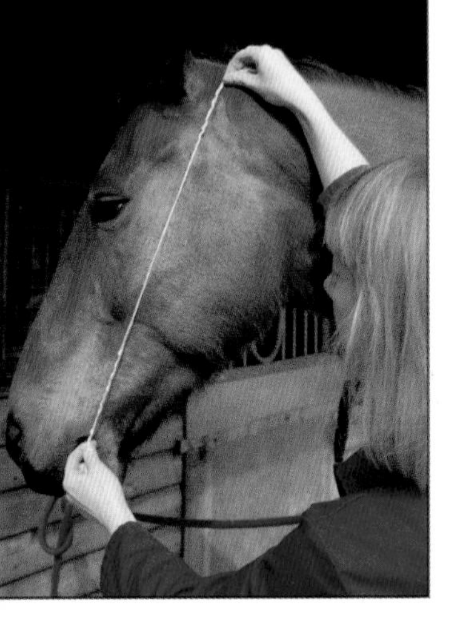

Don't clip the hair from inside the ears, as it forms a protective barrier against insects and seeds from a haynet, if used. Just neaten the edges of the ears by gently folding each one in your hand and running a small pair of quiet clippers down the edges. If your horse won't tolerate this, use a pair of trimming scissors with rounded ends to achieve a similar neat effect.

Legs should be trimmed to complement your horse's conformation, making the most of the good points and minimizing the bad ones. If you want to remove coarse hair from the back of his legs, use coarse or medium blades. Using the clippers in a downward direction, following the lie of the hair, gives a subtle effect and using them against the direction in which the coat grows gives a sharper one.

Some horses object to having their legs clipped, so be careful. If he is reluctant to have a front leg clipped, asking your helper to pick up and hold the other one may persuade him to stand quietly. If he dislikes the clippers on a hindleg, picking up the front leg on the same side may solve the problem. Both you and your helper need to be patient and aware of the horse's reactions.

Fetlock hair can be trimmed off either with clippers or scissors. The easiest way to get a natural look when using the scissors method is to hold a comb against the fetlock and trim the hair that protrudes through the teeth. Change the angle of the comb as you work around the joint.

If at any time during the clipping procedure you feel that you are in danger of getting hurt, or that your horse is frightened rather than wary, stop immediately. Rather than risk injury to either of you, ask your vet's advice about using a mild sedative to break through the fear barrier or call in a professional groom who knows how to deal with difficult horses without using force.

Once you've finished clipping your horse, give him a thorough groom to remove all the last loose hairs, and use appropriate blankets to keep him warm. Be prepared for him to be a bit more lively the first time you ride him, as he'll feel the cold and wind more. Don't let him get chilled—if necessary, use an exercise blanket to keep the muscles on his back, loins, and hindquarters warm when doing slow hacks or when warming up and cooling down before schooling or competing.

On areas where the hair has been taken off and tack rests, the horse's skin will be more susceptible to rubs from girths and reins—particularly reins with rubber hand grips. If you find rubs on his side where your legs rest, seams or zips on boots or chaps may be to blame. Never ride a horse who has a girth gall (rub) as you will cause further discomfort and infection.

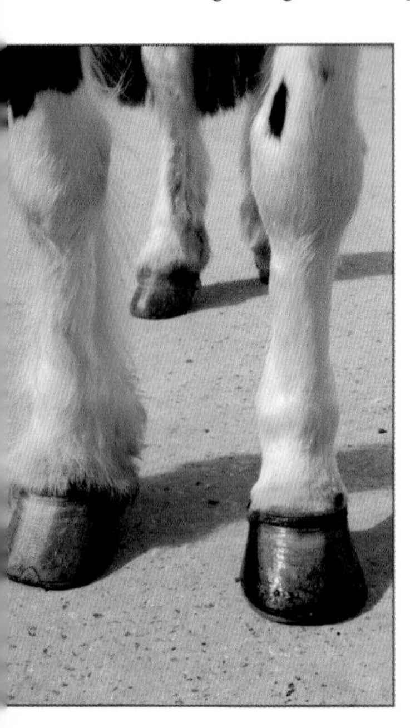

An exercise rug keeps a clipped horse warm.

Trimming and hogging

If you don't want to clip your horse, you can smarten him up considerably with a bit of trimming, using either clippers or a combination of clippers and scissors and comb. Trimming the hair under the jawline will transform a billy goat into a handsome horse once more and trimming his legs will also give a neater appearance.

Cob type horses are often presented with roached manes, where all the mane and forelock hair is clipped off. Many people believe it shows off their workmanlike heads and necks and it is usually obligatory for show cobs to have hogged manes. However, if you don't want to show, it is a matter of choice.

The only effective way to hog a mane is to use clippers in three stages. First, run the blades up the center, then clip each of the sides. When hogging a long mane for the first time, use clippers or scissors to take off most of the excess hair before finishing as above.

If your horse or pony is a recognized breed or type and you want to compete in show classes, check breed or show society regulations before you start trimming. With some breeds, no trimming is allowed and with others, it is limited.

Should you trim off the whiskers on your horse's muzzle? That's a controversial area and very much a personal decision. Some people believe it is essential to provide a clean appearance while others say that whiskers are part of the horse's sensory equipment and as such, should be left alone. Certainly, you should not trim eyelashes or feelers around the eyes.

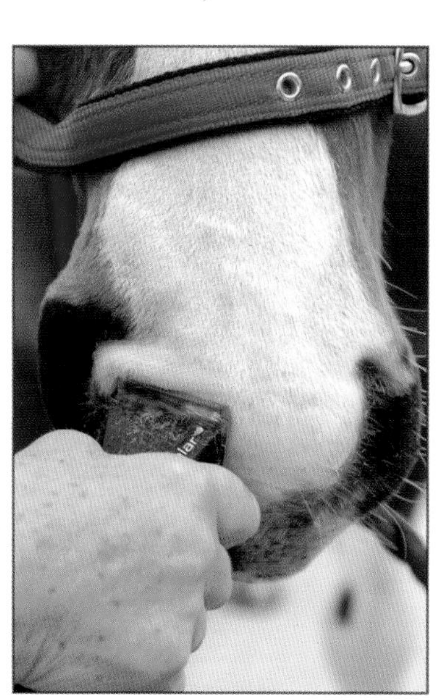

Mane Attractions and Tail Endings

To make the most of your horse's appearance, you may also want to shorten and shape his mane and tail. Again, before doing this you should check breed regulations as some animals are shown with full manes and tails—or at least, with nothing other than some very discreet tidying.

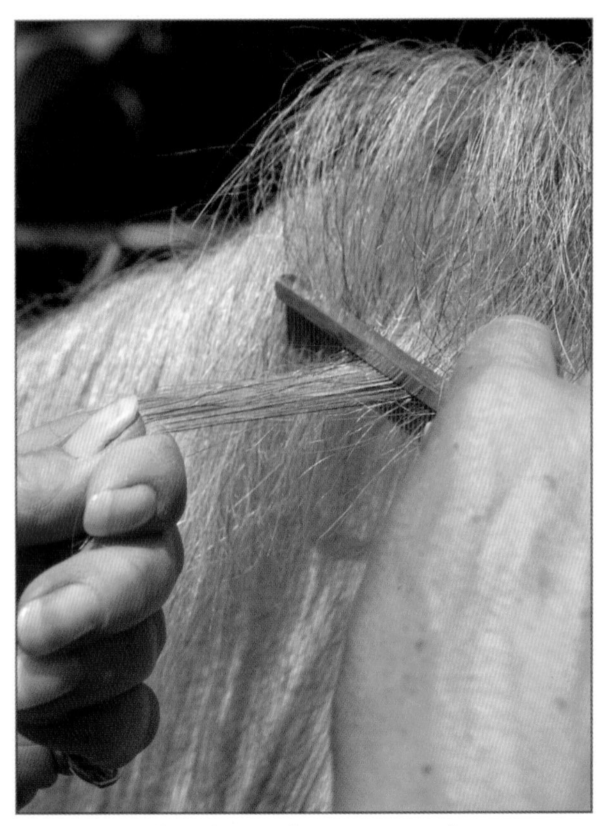
Pulling is the traditional way to shape a mane or tail.

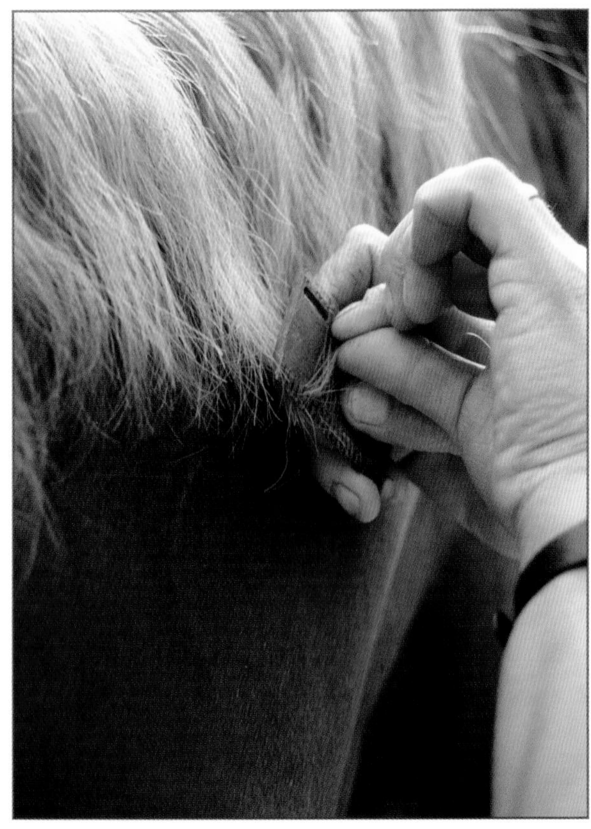
Neaten the mane edge with an old clipper blade.

If you want to plait (braid) your horse or pony's mane in the conventional way for competitions, with a row of plaits up his neck and the forelock made into another plait, you will need to shorten and shape his mane so it is an even 4 inches to 5 inches (10cm to 12.5cm) long and not too thick. However, if you prefer or need to leave his mane long and full, there are plaiting techniques that can be used to keep it out of the way and/or looking nice.

The traditional way to shape manes and tails is by pulling, which is literally pulling out a few hairs at a time. Most horses don't mind having their manes and even tails pulled if it's done considerately and the occasional one even seems to enjoy it. However, pulling tails is not always the best or safest method and if your horse objects, use the alternative technique explained later on.

Always try and pull a mane or tail when the horse is warm, as the pores of the skin are open and the hair comes out more easily. After exercise is usually a good time. When you're pulling a mane or tail that needs a lot of work, don't try and do it all in one go, or the horse might become bored and resentful; if you spread the process over several days, he will be more amenable.

For manes, use a metal pulling comb with short teeth. This can also be used on tails, though some professional grooms prefer to use a nylon comb cut to a suitable length. Horsehair can inflict painful cuts on the finger that applies leverage, so you might want to protect the joint with adhesive bandages.

Only pull hair from underneath the mane, because if you pull it from on top the hair will form a fringe that sticks up as it grows back. Take hold of a few hairs, back comb the top ones out of the way, and make a quick, sharp pull. Wrapping the hairs round the comb before you pull may make it easier.

Some people like to pull one section at a time to the required length and thickness, but it's often easier to keep

A shaping rake is the easiest, safest, and kindest way to shape and trim a horse's tail.

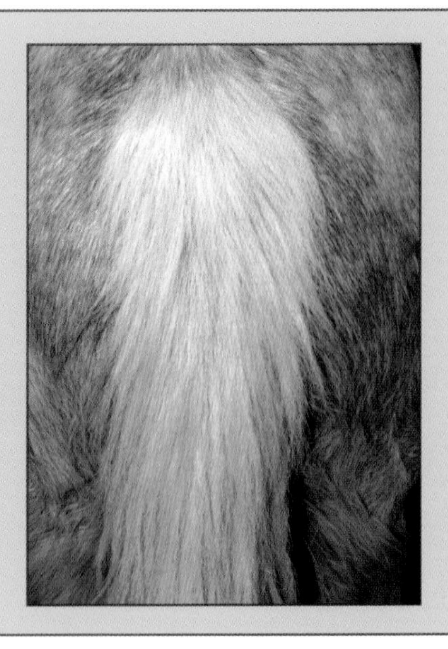

Scissors can also be used to neaten a mane.

working along the mane in stages. When you've reduced the bulk, you might want to neaten the bottom edge with an old clipper blade or scissors.

To use a clipper blade, back comb the mane as is done for pulling and press down with the blade on the hairs underneath. Alternatively, comb through the mane and snip the ends off a few hairs at a time, holding the scissors at an angle. Work up the mane, combing down before every snip so that you get a natural look and not a blunt edge. These techniques can be used when a mane is so fine it just needs shortening rather than thinning.

Occasionally a horse will object to having his mane pulled, even when it's done considerately. The easiest way to overcome this is to use a shaping rake, originally developed for the dog grooming industry. Comb through from underneath and you'll soon thin out the mane. You can then neaten the edges as above.

The traditional way to shape a tail is, again, to pull it. If you're happy to do so and—more importantly—your horse doesn't object and show his discomfort by kicking, first decide how much hair you need to take out. The easiest way is to put your hand on top of the dock and hold it out at the angle at which it is carried as the horse moves; as a guideline, hair which falls away at each side of the dock will need to be pulled.

Pull as before, taking hair from the sides. Try not to pull from the center, or the hair will grow back like an unruly brush. However, if you are dealing with a horse who has a very thick dock and tail, such as a cob, you may need to take a little from the center. In this case, back comb the top hair out of the way and remove some from underneath.

Want to make life easier for you and your horse? Then use a shaping rake and get the appearance of a pulled tail much more quickly and easily, without causing any discomfort to your horse. Simply comb down the sides of the tails in quick, sharp strokes until you have the desired shape.

Whatever technique you use, it's easier to get a uniform appearance if you shape both sides at the same time rather than completing one, then starting on the other. Once you've achieved the desired result, it helps to keep the tail in shape if you apply an elastic tail bandage regularly, as this will train the hair to lie flat.

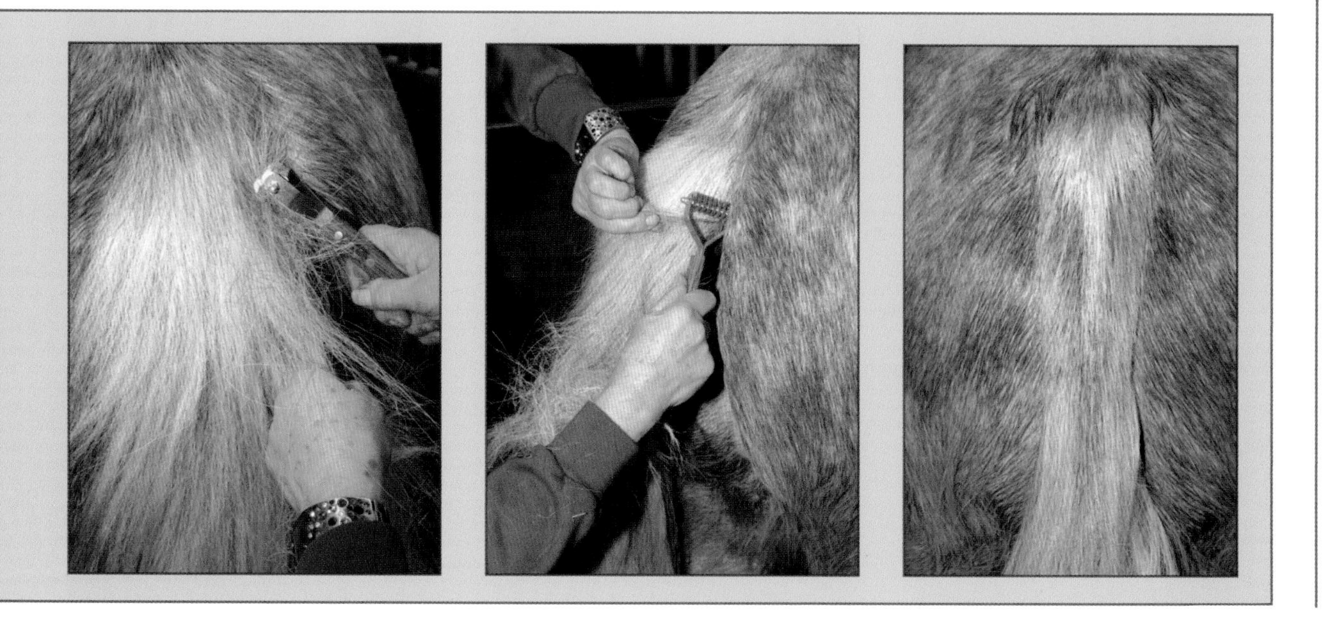

Perfect plaits

Plaiting a mane makes your horse look smart for competitions. However, if you are competing in show classes, check what's correct for your horse or pony's breed, as native ponies and purebred Arabs should be shown unplaited.

Conventional plaits are set in a row along the horse's neck, with another made from the forelock. Traditionally, there should be an odd number along the neck. The old grooms used to stipulate seven or nine, but today, rules are not so strict. However, too many plaits look fussy and you shouldn't need to make more than eleven.

Special decorative techniques are traditionally used for heavy horses.

To keep plaits really secure and smart, you'll need to secure them with stitches. If you're in a hurry or practicing to see what suits your horse or pony, use rubber plaiting bands the same color as the mane.

You'll need the mane to be of even length and thickness, as explained above. It should be clean, but not newly washed, or you'll find the hair is too slippery to grip. Divide it into equal sections and fasten each one with a rubber band.

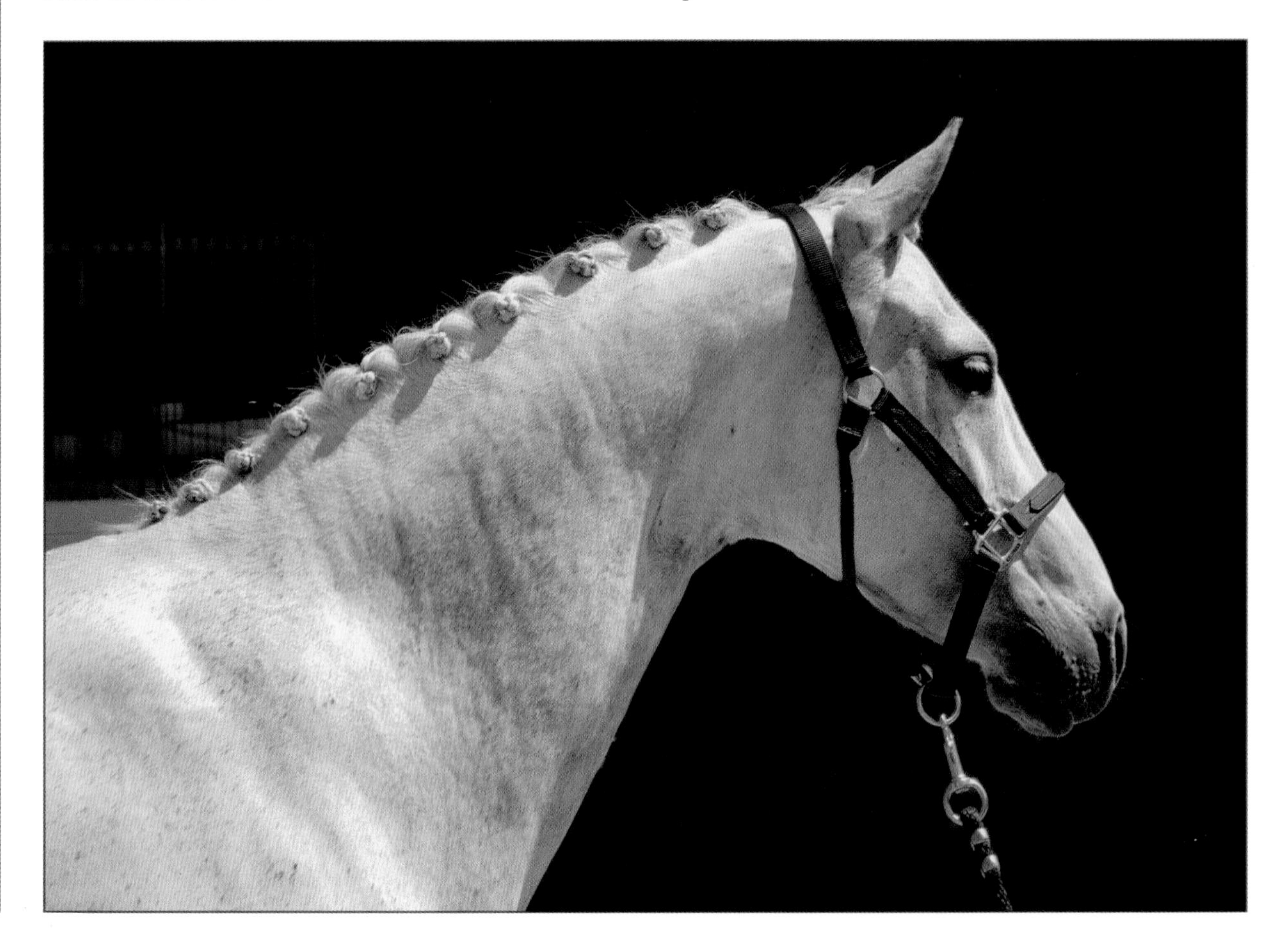

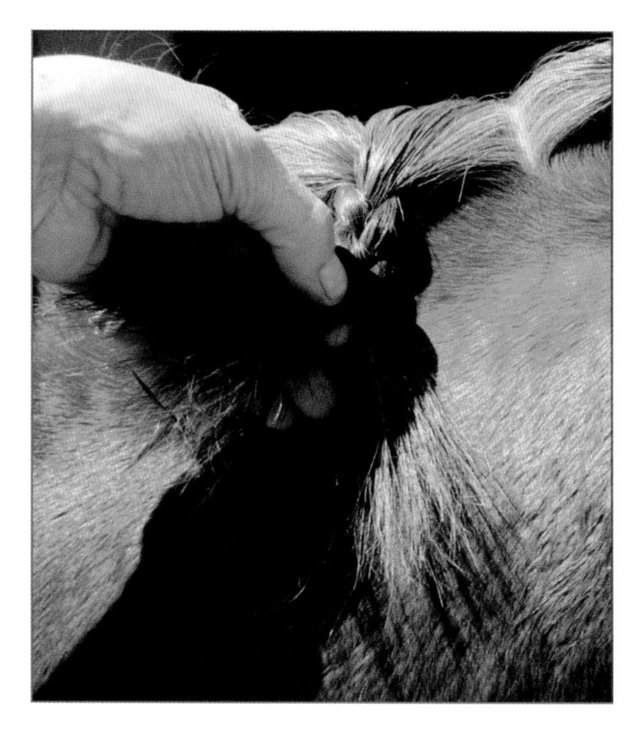

• Dampening each section with water or a spray-on plaiting product makes it easier. Divide it into three equal parts and plait down, keeping the braid tight.

• Plait as far down as possible and secure the end with a needle and thread. Turning over the ends and wrapping the thread around will help to keep the finished plait neat, as the loose ends won't stick up from the base.

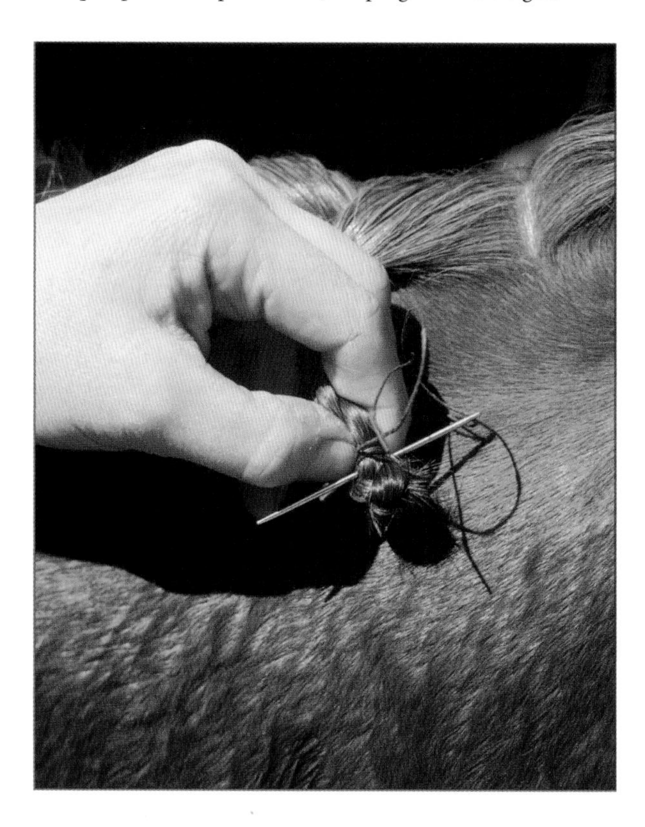

• Next, push the needle through the underside of the plait at the base so that the plait folds in half. Roll up the plait and secure with two or three stitches.

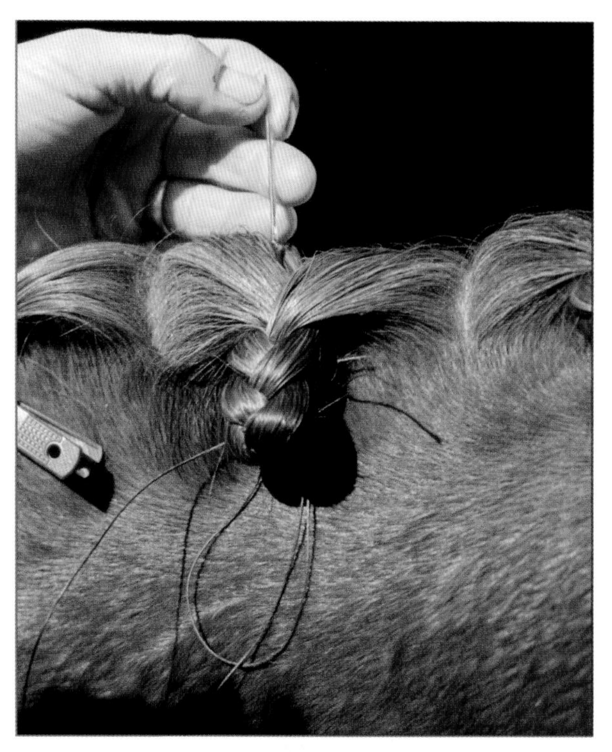

• Carry on up the neck and follow the same process to make a forelock plait. Practice really does make perfect, so don't worry if you don't get it right the first time.

To shorten the process, plait down as before and secure the end with a rubber band. Roll up the plait to the neck and use a second band to hold it in place.

If your horse has a long mane and you want to keep it out of the way, use a running plait. This is sometimes called an Arab plait, as it is often used on Arab racehorses.

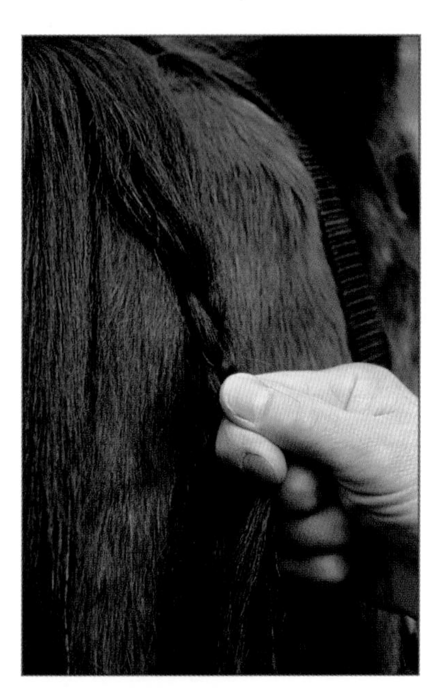

- Starting near the ears, take a small section of hair and start to plait—but each time you pass the left hand section over the center one, take in a small piece of mane. Don't pull the plait tight, as you would with conventional plaits, but let the mane fall naturally.

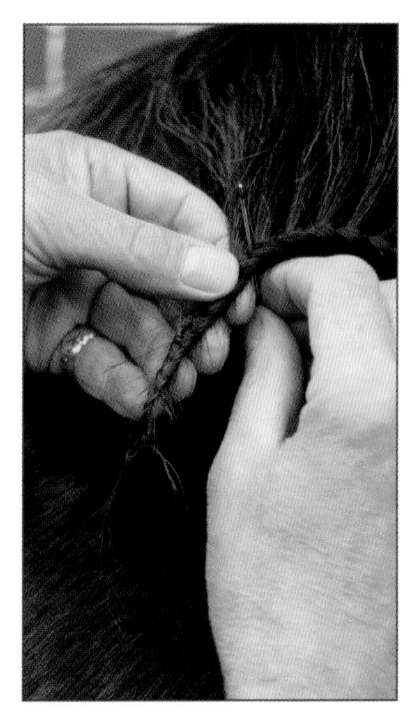

- You'll find that as you progress, the plait will curve around and you'll have a long plait running along the bottom edge of the mane. When there is no more hair to take in, plait the hair that is left, fasten the ends, fold them over, and stitch securely.

If you don't want to pull or shape a tail, you can plait it for special occasions. This is more time-consuming than plaiting a mane and takes a lot of practice, but looks very attractive when done well. You can only plait a full tail with long hair at the sides, so if your horse has a shaped tail that is growing out, you'll have to be patient. Make sure the tail is clean but that the hair isn't too slippery to grip.

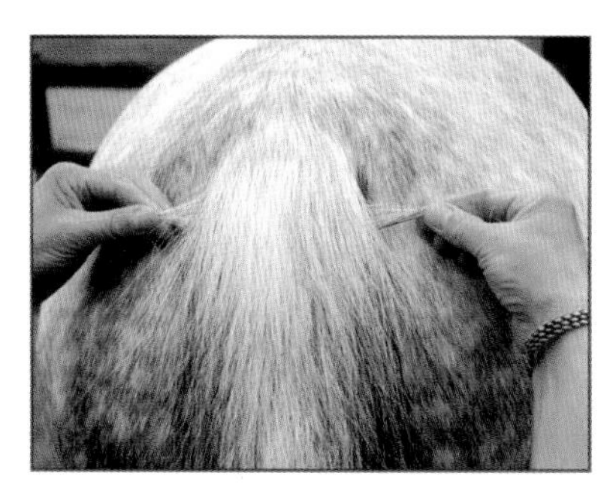

- Start as close to the top of the tail as you can, taking a few hairs from each side. Cross them over and take a third section of hair from the side, just below the first one.

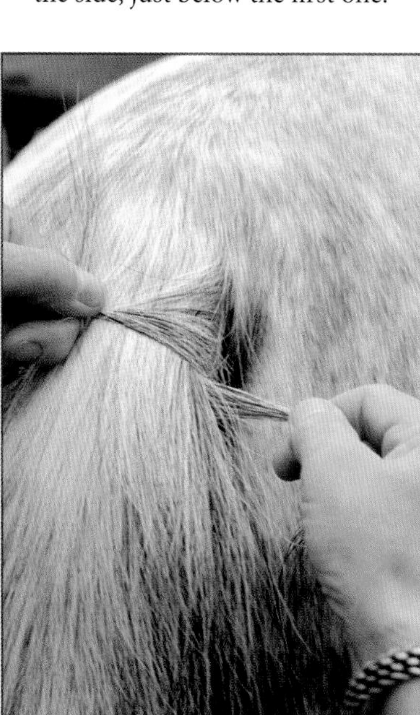

- Now plait down, taking in a few hairs from each side as you pass the sides over the center piece.

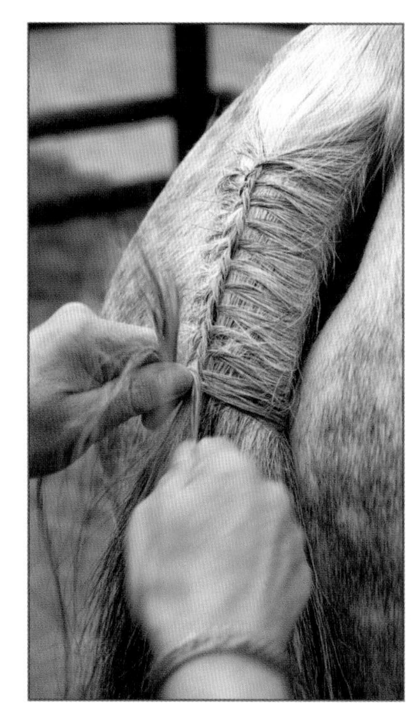

- The key to creating a clean tail plait is to keep your plaiting tight and to take in just a few hairs each time from the sides. This helps to keep the side sections taut and flat.

- When your center plait reaches about two-thirds of the way down the dock, stop taking in any more hairs from the side, but carry on braiding until you reach the end of the plait.

- Fasten the ends neatly with a needle and thread, as with mane plaits, then double up and secure the end.

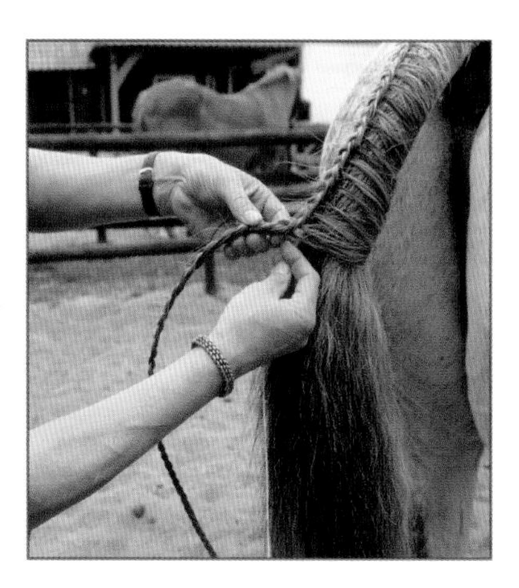

- If you put on a tail bandage when traveling, undo the tape and unwrap it carefully—don't slide it down, or you could ruin all your hard work.

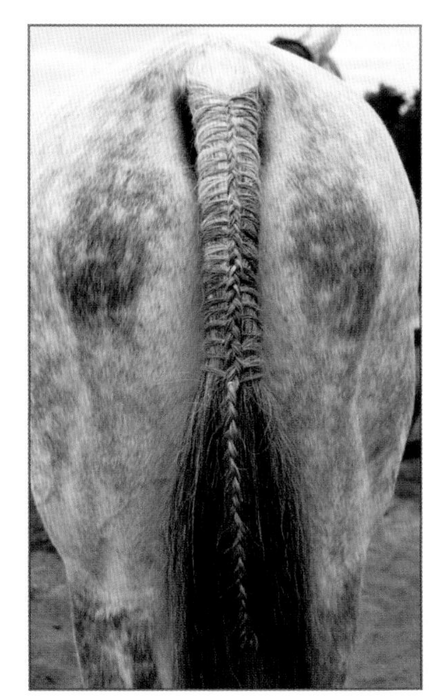

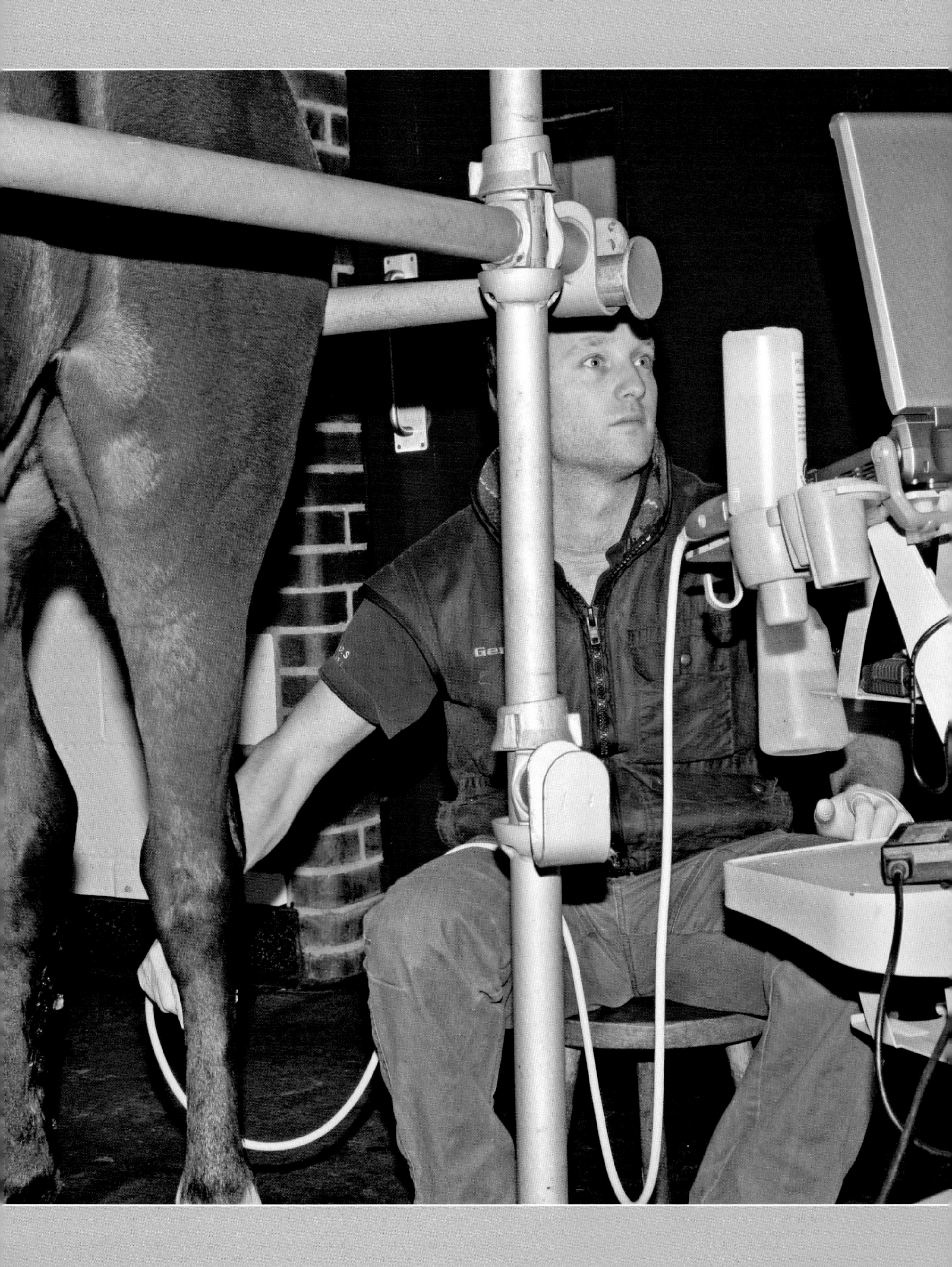

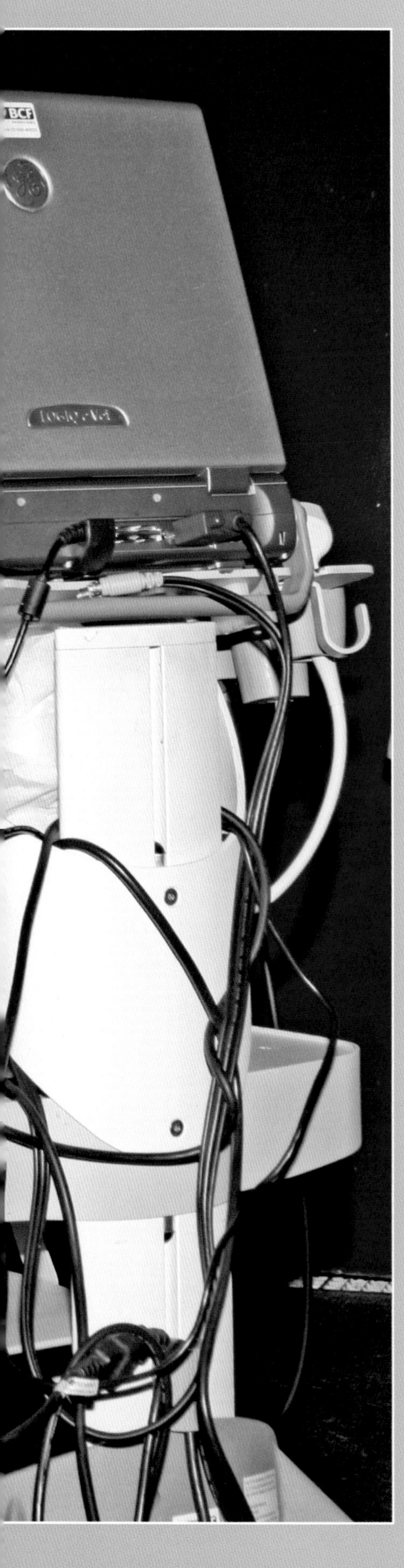

PART 4

HEALTH AND FIRST AID

Keeping a horse healthy is perhaps the biggest responsibility of any owner and is influenced by everything you do—from keeping him in a suitable environment to making sure that his tack fits. Some things will be governed by good practice and common sense, while others will rely on professionals such as vets and farriers.

As far as an owner's responsibilities go, health care comes into three categories: regular preventive care, such as worming; routine care, such as farriery; and first aid. You also need to be observant and to know how to recognize when your horse may have a problem that needs professional attention.

Although the ultimate responsibility for your horse's welfare rests on you, you'll need a backup team of professionals: your vet, farrier, equine dental technician (if your vet does not specialize in dental care), and saddle fitter are all key members of his care team. There may also be times when you will call in qualified practitioners such as physiotherapists and chiropractors, who will work under veterinary approval.

Everyone you trust your horse to must hold a recognized qualification; unfortunately, there are a lot of people who call themselves by terms such as horse whisperers or therapists who do not. While many may be knowledgeable and have valuable skills, there may be cases where at best, you are wasting your money and at worst, putting your horse's health at risk. Many practitioners will only see or work on a horse with your vet's knowledge, which has to be a good thing.

Your farrier is an important part of your team.

Good relations

Hopefully, you'll rarely need to call out your vet except for routine preventive measures such as vaccination. It's always useful and reassuring if you're able to see the same vet each time, so that you get to know each other and the vet gets to know your horse. If you use a large veterinary practice employing several vets, this may not be possible – but even so, you should feel that you can communicate well.

If you don't understand an explanation, ask for a simpler one. If you're worried about your horse but not sure whether he needs veterinary treatment, contact your vet and explain your concerns. A good vet would rather make a visit and find that, for instance, a minor wound doesn't need stitching, than have to deal with an infected and potentially serious one because an owner didn't seek advice.

It's also important to get veterinary advice as a first step in solving a problem, rather than trying to treat it yourself. For instance, if your horse seems stiff in his back or limbs, start by asking your vet to examine him rather than assuming the horse has arthritis and buying a nutritional supplement or magnetic therapy boots which are claimed to help. If the problem is arthritis, it may well be that nutritional support or magnetic therapy will help—but if the cause is an injury because, unknown to you, the horse laid down and got cast in his stable (laid down or rolled over near a wall and can't push himself back up) you won't be helping him or your bank balance.

The spread of information, especially via the internet, means that horse owners are more likely to do research and ask professionals questions. That's a good thing and everyone you deal with should be happy to explain things to you, but remember that there is good and bad information out there. The bottom line is that you should be able to communicate and establish a good relationship with everyone who helps you care for your horse.

Over the last twenty years there have been huge advances in veterinary science and a wide range of diagnostic equipmet and treatment options are available. However, this can be expensive and in some cases, taking out insurance to cover veterinary fees is worthwhile.

The more practical first aid skills you can acquire before you actually need them, the better equipped you will be to deal with problems if they arise. For instance, many veterinary practices will run courses to teach horse owners how to take temperature, pulse and respiration rates, apply bandages, and so on. Learning from practical demonstrations by a qualified person is the best way and will give you confidence.

Insurance

For many owners, specialist equine insurance—in particular against veterinary fees—gives peace of mind. Over the past twenty years there have been many improvements in veterinary science, and while these mean that problems can be more easily identified and, hopefully, treated, it's frighteningly easy to run up a veterinary bill running into thousands of dollars.

It's also important to have third party insurance in case your horse causes injury or damage to another person or property. This can be arranged through a specialist equine insurer, or is a membership benefit of some equestrian organizations.

Although it may sound obvious, make sure you arrange insurance cover to suit your circumstances and read your insurance policy carefully when it arrives. For instance, you will need to insure your horse to take part in all activities you intend to do. If you decide to take part in an activity for which you aren't insured, and he's injured, you won't be covered under the terms of your policy.

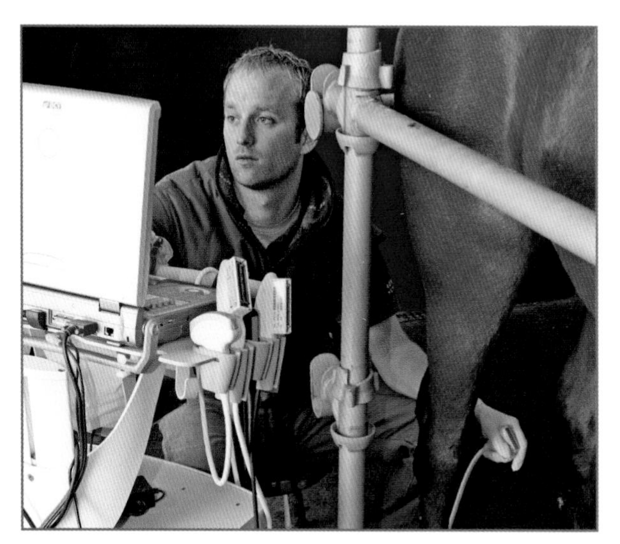

Modern diagnostic and treatment techniques are effective, but can be expensive.

Signs of a healthy horse

When you see and handle a horse every day, you should get to know what is normal for him and what might be a warning sign. For instance, if he's normally waiting with his head over the door for his breakfast, ears pricked in anticipation, a warning bell should sound if you arrive to find him listless and dejected.

There is a simple way of remembering the basics of what to look at when assessing your horse. It's literally as easy as ABC, which stands for appearance, behavior, and condition. Does he look his usual self or, for instance, does he have a runny nose? Is he behaving normally, or is he more listless or grumpy than usual? Is he in good condition, neither too fat nor too thin, as explained in the last section?

Essentials in a first aid kit

Keeping two first aid kits at hand—one on the yard and the other in a horsebox or towing vehicle for whenever you transport your horse—is part of preventive care. While you're at it, remember that there should also be a first aid kit for humans to use!

Although you obviously won't keep it with the rest of your supplies, it's advisable to keep a mobile phone with you on which numbers for your vet and stable should be listed.

Your kits should contain:

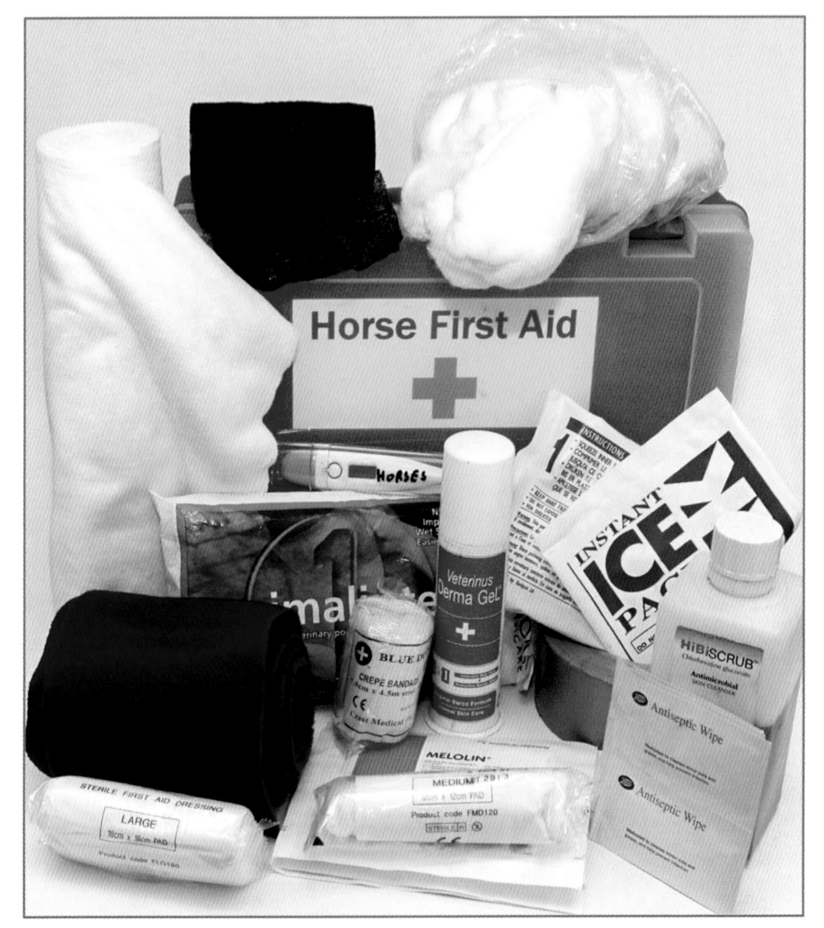

Make sure you have your own first aid kits.

- A card on which you've recorded your vet's telephone numbers, fastened to the inside of the lid
- Digital thermometer
- Antiseptic: your vet will recommend a suitable one
- Antiseptic wipes—if you don't have much room in your travel kit, include these and omit the ordinary antiseptic
- Moist wound gel, marketed under various brand names
- Large roll of cotton wool
- Scissors with curved ends (safer to use than ones with pointed ends)
- Gamgee (cotton wool sandwiched between layers of gauze)
- Poultice for use either wet or dry
- Duct tape for securing poultices or, if preferred, a poultice boot
- Dressings—nonstick ones to cover wounds before bandaging; cotton stretch bandages; cohesive bandages; "cooling" bandages or wraps which can be soaked in cold water or placed in a fridge for a while before use

As part of good management and health checks, you should be able to take your horse's temperature, pulse, and respiration.

- The normal temperature for a healthy adult horse is 100 to 101 degrees F (37 to 38 degrees C).
- The normal pulse of a horse at rest is thirty-five to forty-two beats per minute.
- The normal respiratory rate for a horse at rest is between eight and twenty breaths per minute.

Many top competition yards take their horses' temperatures daily. It only takes a couple of minutes and is good practice if you can fit it into your routine, taking the temperature at the same time each day. This will enable you to establish a base figure for your horse and spot any changes immediately—a slight rise in temperature is often the first sign of a virus and recognizing it will enable you to stop work and avoid stressing him.

Horses' temperatures are always taken in the rectum, so don't put the thermometer in his mouth! Use a basic digital thermometer kept purely for use on horses and grease the end with petroleum jelly. Ask someone to hold the horse, then stand to one side of his hindquarters—never directly behind him, in case he is startled and kicks out. Hold the base of the tail and lift it gently to one side, then insert the thermometer, keeping hold of both.

When the temperature stabilises, remove it gently, take the reading and wipe it clean. It should be cleaned in a

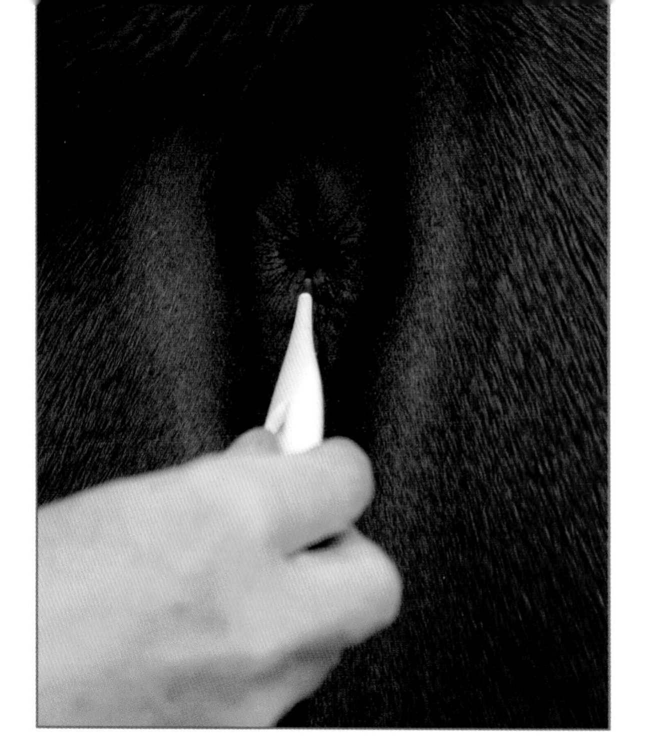

cold water solution of disinfectant before the next use. A temperature higher than 102 degrees F (40 degrees C) is a signal to get veterinary advice. A very low temperature is less common but is also an indicator that something is wrong.

The easiest way to measure the pulse is to feel the facial artery where it passes under the jaw. You can also feel for the heartbeat on the left hand side of the body, just behind the elbow.

To measure the respiratory rate, stand to the rear and slightly to one side of your horse and count the number of breaths. A single breath comprises of both inhalation and exhalation.

There are many factors that affect pulse and respiration. For instance, a horse who has just worked reasonably hard or who is excited will have higher readings, which is why the base rates are taken at rest. Never ignore signs of respiratory distress, such as heaving flanks and flared nostrils at rest.

Other indicators of health to look for include:

- **A Healthy Appetite**: If your horse or pony lives to eat, going off his food can be one of the first signs that he isn't feeling too good. Check water intake, too—while work and weather conditions obviously affect the amount horses drink, marked thirst or reluctance to drink should not be ignored. Dehydration is as serious in horses as it is in people, which is why one of the golden rules of feeding is that clean, fresh water should always be available.
- **Droppings**: Loose droppings or, more seriously, marked diarrhea are a sign that his digestive balance is disturbed and/or that he is ill. However, most horses pass looser droppings if they are excited or nervous. A reduction in the usual number of droppings may indicate constipation or colic.
- **Urine**: The horse should not strain to pass urine and it should have a yellow tinge. If he's having trouble or the urine has a red tinge, call your vet. A sudden increase in the frequency of urination should also be investigated.
- **Mucous Membranes**: The membranes around the eyes and on the gums should be salmon pink, though occasionally you'll find a horse that has black pigment in his gums. When a horse is ill or in shock, his gums may turn a darker colour. Discolored membranes around the eyes are also a warning sign; for instance, if they take on a yellow tinge, it may be a sign of jaundice.
- **Skin**: The horse's skin should feel supple and be easy to move and when you pinch a fold of skin, it should go back into shape immediately you let go. If it remains raised, the horse is seriously dehydrated and needs immediate veterinary attention.
- **Digital Pulse**: The digital pulse is felt where an artery runs over the fetlock and if the horse has a foot problem such as an abscess, is usually raised. If you can't feel it, ask your farrier or vet to show you.

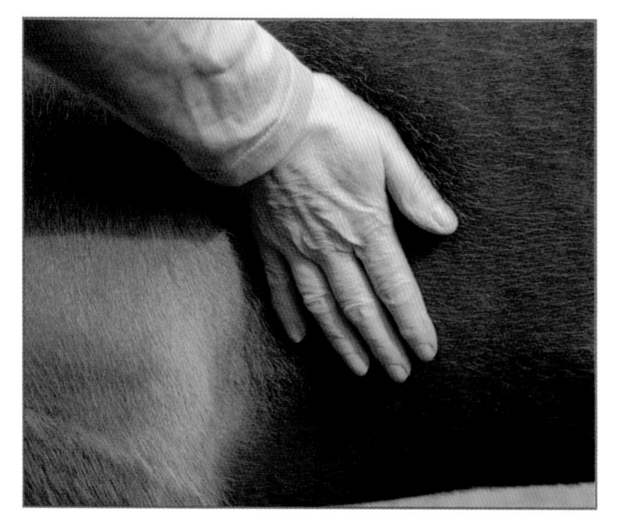

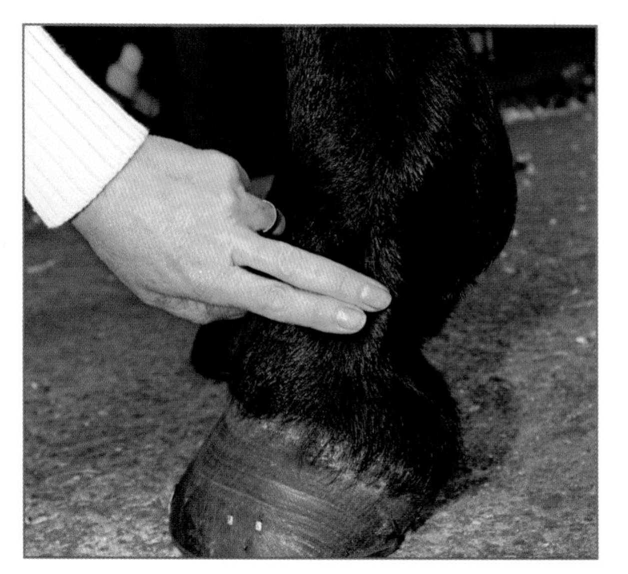

Feeling for the heartbeat and (right) for the digital pulse.

Routine preventive care

Good hygiene practice, vaccination, worming, and regular dental care are part of the preventive care regime that will help keep your horse healthy. Your vet will advise you on appropriate measures, but the information here gives essential guidelines.

Stay Safe

Good hygiene practice is an effective way of preventing the spread of contagious diseases. This includes making sure every horse has his own tack, clothing, grooming kit, and feed buckets rather than sharing them with others.

If you're worried that your horse is showing signs of illness, such as a raised temperature, coughing, or a runny nose—often coupled with lethargy—ask your vet to check him. It's far better to spot problems early than to wait until several horses on the yard are affected.

It's often recommended that every new horse who arrives on a yard should be isolated—able to see other horses but not in contact with them—until it's clear there are no signs of disease. In practice, this often isn't feasible. When it is, it's only worth doing if everyone who comes into contact with him stays clear of other horses. Also, remember that horses are herd animals who need the company of their own kind. Moving to a new home is always stressful and if a horse is quarantined for two or three weeks, he must be able to see others even though he is not able to make physical contact with them.

If there is an outbreak of disease at your stable or in your area, your vet should be able to advise you on strict hygiene and cleanup measures.

Left: Every horse must have his own grooming kit. **Below:** A horse who needs to be isolated should still be able to see others.

Vaccination

Horses in different parts of the world are vulnerable to different diseases, as are their owners, so wherever you live, it's vital to check with your vet on the latest information and take any preventive measures recommended. It's easy to be a scaremonger, but diseases such as West Nile Virus and African Horse Sickness have spread rapidly and now that international transport of horses has become commonplace, vets are alert to the growing risks. There are also diseases such as strangles and equine grass sickness which are the subject of international research.

So what do you need to protect your horse from as standard? The basics are tetanus and equine influenza and, where appropriate and on veterinary advice, equine herpes virus and equine viral arteritis (EVA).

Tetanus, colloquially known as lockjaw, is caused when spores of a particular bacteria spread within a wound and produce toxins. Vaccination is cheap and simple and every equine should be protected. Incidentally, so should you, so check with your doctor.

Equine flu is debilitating and spreads like wildfire amongst unvaccinated horses. A primary course of two injections is followed by boosters; your vet will advise you on recommended intervals. Apart from the necessity of protecting your horse's health, they are mandatory under the rules of some equine disciplines' ruling bodies.

Vaccinations against equine herpes virus and EVA are not given routinely, but your vet will advise you if and when they are necessary.

Worming

Every horse has an intestinal parasite burden, but a good worming program tailored on veterinary advice to your circumstances will enable you to keep your horse healthy. Worms are not a minor problem—at best, they mean your horse may have a dull coat, feel lethargic, may lose weight, and at worst, they can lead to potentially fatal conditions such as colic (see emergencies section.) However, a horse can look perfectly well and still have an unacceptable worm burden, so it isn't something you can leave to chance.

Know Your Parasites!

The commonest worms are small redworms (*Cyathastomes*). They pose a particular risk because they can hibernate in the gut wall and become encysted, emerging in the spring and damaging the stomach lining as they do so. This is a common cause of colic.

Large redworms (*Strongylus vulgaris*) and large roundworms (*Parascaris equorum*) are equally harmful and if they emerge in your horse's droppings after worming, you won't miss them—they can reach up to 16 inches (40cm) long. As they pass through the body, they travel through the lungs and can cause respiratory problems.

Tapeworms (*Anoplocephala perfoliata*) are another major cause of colic. They shed rectangular segments which may be seen in droppings, but the fact that you can't see any doesn't mean your horse isn't affected.

Lungworms (*Dictyocaulus arnfieldi*) used to be a problem when horses were grazed with donkeys, or in fields which donkeys had occupied. They live in the lungs and, not surprisingly, cause coughing. Modern drugs will kill lungworms, but it's important to worm horses and donkeys.

Pinworms (*Oxyuris equi*) are thought not to cause internal problems but will cause itching and discomfort round the anal area. They lay their eggs around the anus and if seen, should be wiped off.

There is one other common parasite that isn't a worm, but the larva of a fly. Bots (*Gasterophilus intestinalis*) can first be spotted in the egg stage on the horse's coat—they look like small yellowish seeds on the hair and are often seen on the legs. Use a bot knife to scrape them off the coat when you see them.

At one time, the accepted approach to worming was to dose every animal at regular intervals with drugs designed to target particular parasites at particular stages of growth. Now, though, a strategic worming approach is recommended when possible.

This involves taking small samples of the horse's dung at regular intervals and sending it for laboratory analysis. By counting the number of worm eggs, it can be determined whether the horse needs worming at that time. However, this will not show whether or not the horse has tapeworms, though a blood sample can be taken by your vet and sent for analysis to determine this.

Depending on the result of regular fecal egg counts, your vet will be able to tailor a program for each individual horse. There will still be times when you need to worm, but the drugs you use can be targeted and once you have the worm burden under control, you are unlikely to have to do it as often. This is not only more environmentally friendly, but will save you money and lessen the risk of parasites developing resistance to certain drugs.

The other vital way of minimizing the parasite burden is good land management. In particular, picking up droppings as regularly as possible—preferably daily—will reduce the number of worm eggs on the pasture and thus the likelihood of them being taken in by the horse as he grazes. It's also important not to overstock fields, as the more horses are grazed there, the more likely it is that they will be forced to eat grass near droppings and thus ingest worm eggs. The worm problem is literally a vicious cycle—the eggs go in one end and out the other!

Wormers in paste formulation are syringed into the horse's mouth.

If you buy a horse, it's sensible to get a fecal egg count done straight away, even if the seller assures you that a stringent routine has been followed. Depending on the results, your vet will then be able to advise you whether the horse needs worming right away and if so, what to use.

Worming drugs are available in a variety of formulations—paste, liquid, granules, or tablets. As they can now be made in palatable forms, many horses will take them quite happily if they are mixed in with a feed; adding a little molasses or honey will make them even more palatable. However, some horses will take the attitude of "There's something I'm not sure about in my dinner and I'm not going to eat it," and if the wormer is in a syringe container, you will have to put it directly into the horse's mouth.

On large livery/boarding yards it may be standard practice that every new horse is wormed on arrival and that all horses are wormed by the yard owner at synchronized intervals. Unfortunately, you may be conscientious, but other owners may be less so.

The golden rules for keeping your horse as free as possible from parasites are:

- Pick up droppings as often as possible, preferably every day and at least twice a week. Harrowing isn't a good idea—it spreads the droppings and if done on a hot, sunny day some of the worm eggs will be killed by ultraviolet light. However, usually all that happens is that by spreading the droppings, you spread the worm eggs over a larger area and make it more likely that horses will ingest them when grazing.
- Don't overstock fields.
- Worm according to the horse's body weight and if you don't know it, use a weigh tape to give you a reasonably accurate idea. If you give too small a dose, your worm control program will be ineffective.
- When possible, use a strategic worming program.
- Get your vet's advice on using the right worming drugs at the right times to make sure you cover all the different types of parasites at various stages of their development.

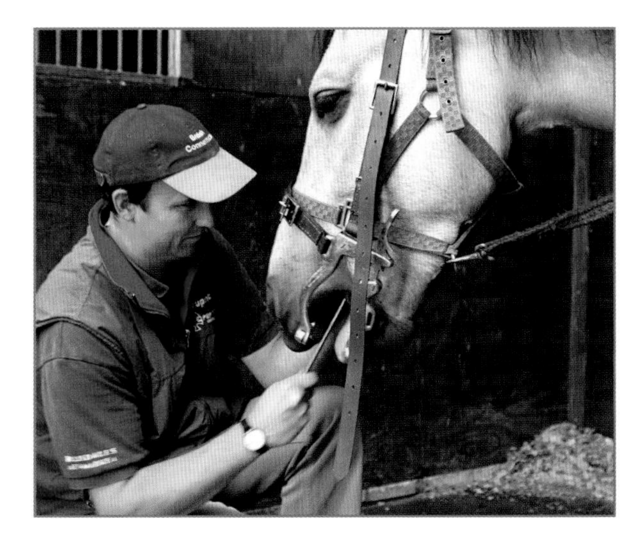

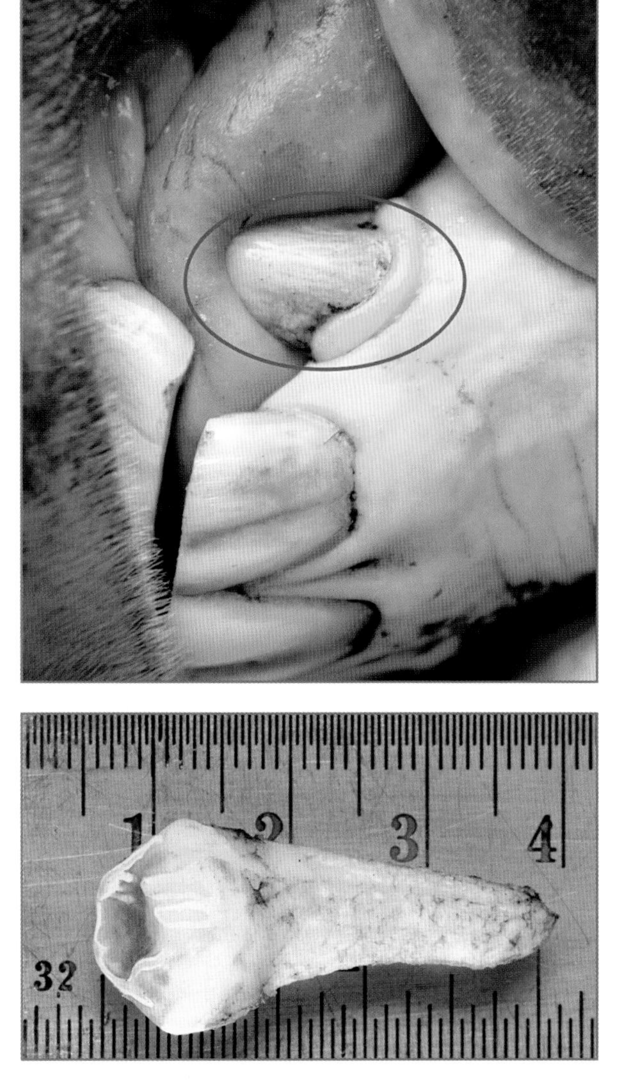

Open Wide

Dental care is as important for horses as it is for humans. Making sure that their teeth and mouths are in good condition affects not only the way they eat and therefore the value they get from their food, but also their performance. A horse who is uncomfortable in his mouth will be even more uncomfortable when asked to answer the rider's signals via the bit.

While our teeth erupt and then stop at a certain point, horses' teeth are grown throughout their lives. They are worn down by eating—in particular, by grazing—but because the diet of the domesticated horse is different from that of those in the wild, the rate of wear is not sufficient and he needs regular rasping to enable him to chew properly.

All adult horses have six front teeth (incisors) in each jaw; twelve cheek teeth (premolars) and twelve molars. Male horses also have four tushes, one at each side of both jaws, which are the remains of canine teeth. You may occasionally find these in mares, but they are usually smaller. In addition, horses of either gender may develop very small, shallow-rooted teeth called wolf teeth, which are vestigial pre-molars.

If you look at the shape of a horse's head, you'll see that the lower jaw is narrower than the upper one. As a result, when he grinds his food, the outside edges of the upper cheek teeth and the inside edges of their lower counterparts gradually develop sharp edges. These can cause rubs and lacerations on the inner cheeks and tongue and, because they interfere with the digestive process, can lead to problems such as colic. Once they have been floated or rasped the horse should be comfortable again.

A vet or suitably qualified person should see every horse or pony regularly to check for problems and carry out remedial work. This applies whether or not the animal is being ridden, as retired horses and ponies and those used for breeding purposes are just as likely to develop problems as those who are ridden or driven.

Different countries have different regulations (or, more seriously, lack of them). Vets can carry out dental work, though some do not have specialist knowledge and may refer their clients to those who have made it a particular study, or to equine dental technicians.

There are no hard and fast rules about how often a horse should be checked. For some, an annual checkup is sufficient while others may need seeing twice a year. However, if at any time you suspect a problem—perhaps because your horse drops food when eating, tilts his head when ridden, or is generally resistant to the bit—call in your vet or EDT. He or she will then advise you on when the next checkup should be made.

You can also get advice on whether wolf teeth should be removed. They are sited close to where the bit rests in the mouth and if the bit mouthpiece bangs against them, it can cause discomfort. It used to be standard practice to remove them, but many practitioners now have the attitude that "If it ain't broke, don't fix it," and believe that unless they are causing a problem, they should be left alone.

This clench needs to be tapped down to match the others.

Best Foot Forward

Making sure that your horse's feet are in good shape and that if appropriate, he is shod correctly, are part of routine management. However, it is also essential to keep him sound, which is why it is dealt with in this section.

Hoof care is a joint effort between you and your farrier. Your responsibilities are to pick out the horse's feet to remove dirt and anything that may have become lodged, such as stones, and to check the condition of his shoes to make sure that they are secure and that there are no risen clenches. Clenches are the ends of the nails that hold the shoes in place; they are nipped off and tapped over by the farrier, but may sometimes come up again.

You also need to make sure that your horse's bed is kept clean. If he stands on dirty, wet bedding for long periods, he is liable to develop foot infections such as thrush. This is one reason why deep litter bedding is not recommended.

All horses, whether working or not, should be seen by a farrier about every six weeks so that their feet can be checked, trimmed, and—if shod—have shoes replaced or, if they are not too worn, refitted.

Shod or Barefoot?

In recent years there has been a growing move to working horses barefoot, without shoes. Although some maintain that every horse can be kept this way with proper maintenance and management, it is vital that every horse should be looked at as an individual, taking into account his workload, environment, and owner's circumstances.

It's also important to appreciate that trimming is a skilled job. It isn't just a case of taking off excess hoof just as you would cut your fingernails when they grow too long—the way the foot is shaped affects the way the horse moves and the stresses that are imposed on the forelimb. Hooves cannot be looked at in isolation and a trained farrier will take into consideration the horse's limb conformation, the shape of the foot, and the way the horse moves.

In some cases, it's possible to work a horse with shoes only on his front feet. Not only does he naturally put more weight on his forehand, leaving him barefoot behind can be a useful safety measure. If a horse kicks out at another

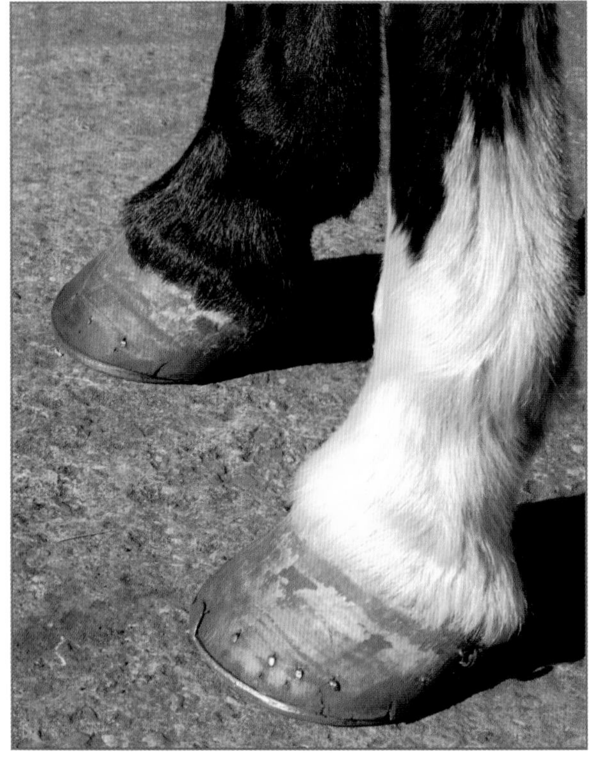

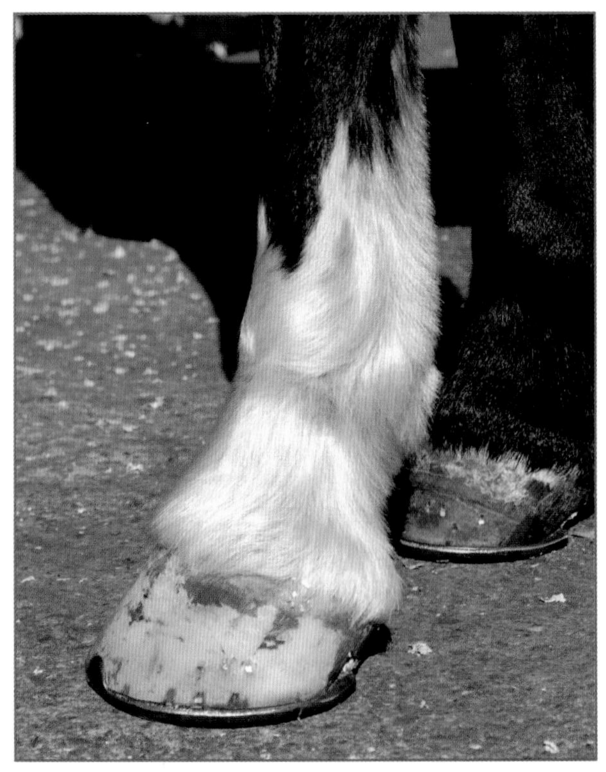

Feet in need of trimming (left) and (right) correctly trimmed and newly shod.

Principles of good farriery

The feet should be trimmed so that from the side, you should see the angle of the pastern running into that of the hoof, forming a straight line from the top of the pastern to the toe. This is called the hoof pastern axis (HPA).

If the horse is shod, the shoes should be of an appropriate weight and size for his feet and overall build and long enough to support his heel. They should be made or altered to match the shape of the correctly trimmed foot rather than the foot being trimmed to match the shape of a shoe.

When a shoe is fitted, it should look as if it's an extension of the foot, with no gaps between the two.

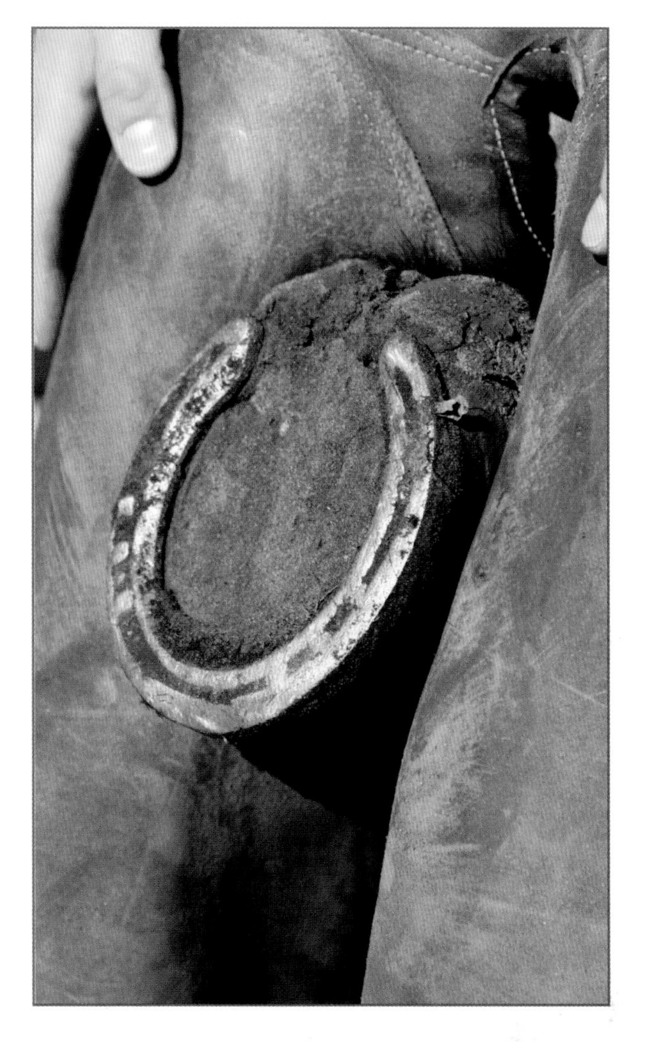

one—or at a person, come to that—a shod foot is likely to do more damage than an unshod one. Talk to your farrier if you want to try this system to see if it will work for your horse and your situation.

At the moment, anyone can, in theory, trim a horse's feet. However, it is strongly recommended that whenever possible, owners use a registered farrier to carry out all hoof care, including trims.

If your horse is shod, it's a good idea to learn how to remove a shoe if it starts to come loose. Your farrier will show you and help you obtain the tools to make it easier, but the picture sequence here will explain the safest way to do it. Don't risk riding a horse with a loose shoe, as there is a danger that the horse will be injured.

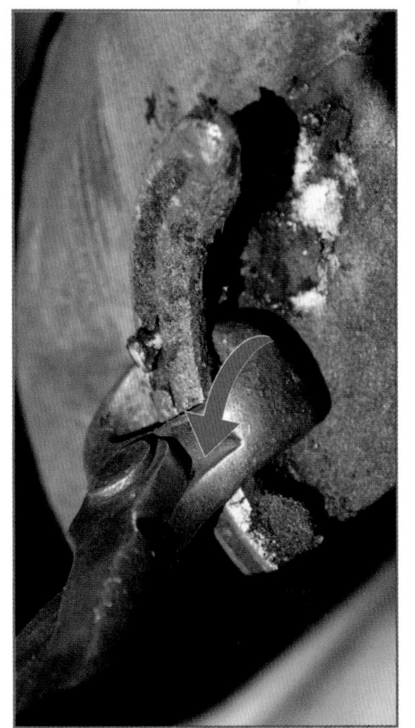

It's an emergency

If you are new to horse owning—or even if you are experienced—it can sometimes be difficult to know when you need to call out your vet. The simple answer is that if you have any doubts about what to do or whether your horse needs veterinary treatment, you should pick up the phone. A good vet will never think that you are wasting his or her time and it is always better to be safe than sorry.

In the following pages, we'll look at some of the conditions and situations you may have to deal with. As a guideline, always call and tell your vet what's happening immediately if you encounter the following situations:

- If you think your horse has a fractured limb, or if he is so lame that he can't bear weight on the affected limb.
- If he becomes suddenly and severely lame.
- If he is recumbent and unable to get up. This is different from a horse being cast.
- If he shows signs of colic.
- If the edges of a wound have pulled apart. If it is a relatively new wound and is clean, it may need stitching. An older wound that has become contaminated with dirt usually can't be stitched, but still needs to be seen and treated by your vet.
- If there is continuous bleeding—in particular, signs of arterial bleeding, when the blood will spurt or pump out rather than flow.
- If you find a deep puncture wound, usually caused by a nail or other foreign body.
- If your horse shows signs of laminitis.
- If he develops continuous diarrhea.
- If one or both of his eyes is swollen, closed, or injured.
- If he shows signs of choking.

It's always worrying or even frightening when someone is ill or injured. In some ways, it can be worse when you're dealing with an animal, because he can't tell you which part of him hurts or how he's feeling. To do your best for him, you need to try and keep calm, assess what has happened, and where appropriate, get help as soon as possible. When you're assessing what has happened, always keep yourself safe. A horse who is in pain or is distressed won't be his normal self, so don't take risks.

If you find your horse in the field with a suspected broken limb or unable to bear weight on one of his limbs, it's better to ring your vet rather than try and move the horse. If the horse has suffered a fracture, your vet will need to apply a support before the horse is taken for treatment.

When you call your vet, keep calm but explain you have an emergency. Listen carefully to anything you are asked and to any instructions you are given before the vet arrives. If you are with your horse in his field, try and contact a friend or fellow owner on the yard so that someone can show the vet exactly where to go.

A badly injured horse will often be in shock and may start shivering with cold. Don't try to move him, but if other people on the yard can be contacted, get someone to bring you a blanket to put over him.

If you think your horse may need to be transported to a veterinary practice or hospital for treatment, try and save time while you are waiting for your vet by organizing transport. In some cases, particularly those involving leg injuries, a horse will find it easier to walk up a gently sloping ramp than a steep ramp. Make sure that whoever is to drive the horse can be relied on to stay calm and drive smoothly.

Common conditions

. . . and new threats

Reading books like this can turn you into a horsey hypochondriac, but while you'll hopefully have to deal with few problems, you need to recognize them and know how to take the appropriate action. You also need to be aware of relatively new threats and, without panicking, be aware of the impact they might have.

African Horse Sickness (AHS)

AHS has been described as the most devastating horse disease on the planet. It is carried and transmitted by midges and sadly, the majority of sufferers die. At present, it is confined mainly to Africa, but vets warn that climate change means it is now theoretically possible for it to reach and thrive elsewhere.

AHS is not contagious, so it cannot be spread from horse to horse. Nor can it be passed from equines to humans. Signs include nasal discharge, excessive salivation, and swelling of the eyes and/or head. While the first two signs are associated with other, much more common conditions such as strangles or equine influenza, it's important that any horse which shows them should be isolated and veterinary advice obtained immediately.

All the preventive measures which apply to sweet itch, such as using effective insect repellent, apply to AHS.

Colic

Colic means any sort of abdominal pain and can range from discomfort to a severe condition that may necessitate surgery. Warning signs include rolling, pawing the ground, looking back at the flanks, and sweating. Often, a horse will show a mixture of these signs and he may also fail to pass droppings.

If you go into your horse's stable and find that there are no droppings or a greatly reduced number—particularly first thing in the morning, when he has been stabled overnight—pay careful attention to how he seems. Take his temperature and if he seems off color and/or his temperature is higher than normal, call your vet for advice.

Equine Rhabdomyolysis Syndrome (ERS)

This may still be referred to by the old colloquial names of azoturia, set-fast, or tying-up. It is similar to muscle cramp in the hindquarters and is just as painful for the horse; signs range from mild discomfort to refusal to move, accompanied by sweating and other signs of pain.

If you suspect the horse has suffered an attack of ERS and are at or near home, put a blanket or your jacket, if available, over his quarters and lead him in gently. Put him in a stable with a thick bed and keep his back and quarters warm with a blanket, preferably one which will leave the layer next to his skin dry by wicking sweat away to the outside. If you are away from home and the horse's movement is restricted, try and keep him warm with anything available and call for help to transport him back.

In a really severe attack, the horse may not be able to move or even remain standing. Call your vet immediately and if you are on a public road, call the police so they can get someone to you to make sure you and other road users remain safe. Again, try and keep his hindquarters warm.

Always call your vet if you suspect your horse has ERS, even if the attack is mild and the horse remains mobile. It is important that blood tests are taken as soon as possible after the attack. Your vet will advise you on management and take further samples to see when the muscle enzymes have returned to normal and the horse can be put back into light work.

Grass Sickness

Colic-like symptoms are sometimes one of the symptoms of grass sickness. A definitive cause has not been isolated, though it is thought to be caused by a soil-borne bacterium called *Clostridium botulinum.*

Grass sickness can be acute or chronic. Sadly, acute grass sickness is always fatal. A horse suffering from it loses weight drastically and quickly and apart from colic-like symptoms, may sweat, show muscle tremors, and have difficulty eating.

Research has shown that although it can affect horses of any age, those aged between five and nine years old are at the highest risk. The danger period for its occurrence is from May through to July, though it is also seen at other times of the year.

In areas where it is prevalent, your vet may advise keeping animals off grass during the danger period or—as there may be a connection with weather—when a temperature of 45–50 degrees F (7–11 degrees C) has been recorded for ten consecutive days. Try and avoid management changes between April and July and feed hay or haylage throughout this period.

Lameness

When a horse becomes lame, it can be a minor or major problem. The problem is that sometimes, it's difficult to tell which scenario you are facing. If the horse can't bear weight on one limb, is lame in walk, if you suspect a fracture, if he is lame in association with a potentially serious wound, or suddenly becomes noticeably lame, it's important to call your vet immediately. There are different schools of thought on how to deal with the horse who shows slight lameness in trot and some people prefer to rest him for a few days to see if the problem resolves.

However, it's safest to call your vet whenever your horse is lame, no matter what the degree. Depending on your description of the problem and any accompanying signs, such as heat or swelling in the limb, your vet will decide whether to make a visit that day or to tell you to rest the horse and report back after a couple of days. Don't ignore heat and swelling, particularly in the tendon area, even if the horse is sound, as the quicker a tendon injury is identified and treated, the better the outcome is likely to be.

Never feel stupid if your vet makes a visit and advises you to rest the horse for a couple of days rest, after which you find that the horse is back to normal. That's a great outcome all around and your vet will be as pleased as you are. On the other hand, if you delay making that call and the horse turns out to have a problem you couldn't identify and which needed urgent attention, you will feel very guilty about possibly compromising your horse's recovery.

Before calling your vet or trotting up the horse in hand, pick out his feet to make sure he doesn't have a stone or other object wedged in one of them. If he does, and goes sound as soon as it is removed, then you really would feel silly.

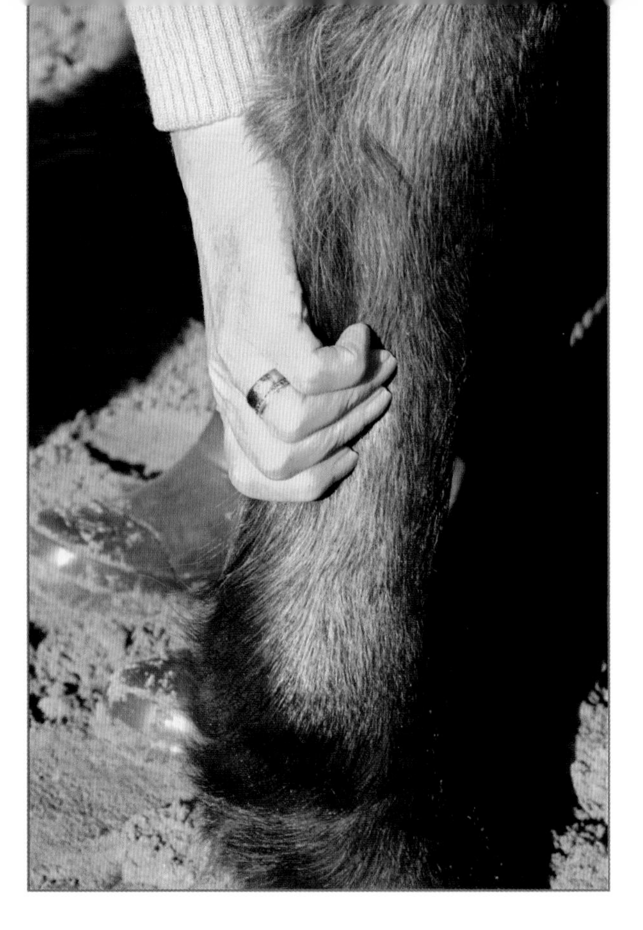

Which Leg?

Unless the horse is not bearing weight on one limb, or has an obvious injury, your first task is to identify which leg is affected. In most cases, a horse will only show lame steps in trot, so you need to either ask someone to trot him up in hand for you, or—if you prefer—trot him up so that someone more experienced, such as a yard manager, can look. He should be trotted away and then back to the observer on level ground, in a safe environment, with the handler keeping a loose lead rope so the horse has freedom to move his head. Cases of slight lameness may show up more clearly when the horse is lunged on a circle, but it may be safer to wait until your vet asks you to do this.

Front limb lameness is easier to identify than that in a hind limb. When a horse is lame in front, he will raise his

head as the lame leg hits the ground and nod down as the sound leg hits the ground. An easy way to remember what happens is to say that he "sinks on the sound leg."

It's easier to identify hind limb lameness when you are watching the horse trot away from you. Watch his hips, as he will lift the hip on the lame side higher in his efforts to avoid taking weight on it as it comes down.

Whether you think the horse is lame in front or behind, it's vital to watch him moving away and towards you to make sure you don't confuse a severe hind limb lameness with a fore limb one. Although a horse won't usually nod his head if he is slightly lame behind, he may do so if the lameness is more painful.

He will try and move his weight forwards to lighten the load on his hind legs and because he moves his legs in diagonal pairs at trot, it's easy to get mixed up. For instance, if he is very lame on his left hind, his head may nod down as the lame hind leg hits the ground. At the same time, his right fore leg hits the ground, so if you think that this is a case of him "sinking on the sound leg," you may mistakenly assume that he is lame on his left fore limb.

The final part of the lameness puzzle is the horse who is lame on two legs rather than one. It may take a vet to identify this, as one set of clues can almost rule out the other. However, a horse who is lame on both front legs—which is the usual scenario—will take shorter, more shuffling steps than normal.

Obvious Causes

By far the greatest number of lameness problems are caused by pain in the foot rather than pain in the limb itself. There can be all sorts of reasons—for instance, the horse may have bruised his sole by picking up a stone, he may have an abscess in his foot which needs draining, or he may be showing signs of laminitis (see later in this section).

In many cases, a foot problem will be accompanied by a stronger than normal digital pulse. This can be felt where an artery passes over the fetlock and in normal circumstances can be quite hard to find. If your horse is lame or reluctant to move and you can feel the digital pulse easily, it's likely that there is a problem in that foot. If there is a strong digital pulse in both front feet, a likely cause is laminitis.

Tendon problems may or may not be accompanied by heat and swelling, but are a good reason for calling your vet rather than deciding to wait and see. If diagnosed and treated early, the outcome may be more successful than if you continue to work the horse and make the injury worse.

Laminitis

Laminitis is a painful condition of the feet that can affect animals of all types. It's often thought of as affecting fat little ponies—and often does—but is equally likely to occur in horses, including Thoroughbreds.

There are several trigger factors, the most common of which is an overload of fructans, a type of carbohydrate in grass and plants. The levels are at their highest in spring and autumn when grass is growing most efficiently, but climate change means that this can happen at other times of year too. Fat animals are at greatest risk because of alterations in their metabolism.

Other trigger factors that have been identified include concussion, a severe infection which leads to toxemia and stress. In many cases, there may be a combination of factors involved. In mild cases, the animal may take shorter strides than normal and in severe ones, he may be unable to walk because of the pain and will try to keep his weight off the affected foot or feet; if all four feet are affected, he will usually want to lie down.

It may seem like a cliché to say that prevention is better than cure, but preventing a horse or pony from becoming overweight is one of the best ways of minimizing the risk of laminitis. This is best achieved through a combination of appropriate diet and a sensible exercise regime. If your horse is already overweight, don't "starve" him even if your intentions are good, or you run the risk of him developing other health problems. Instead, get advice from your vet and/or a good nutritionist about giving him a high fiber, low calorie diet supplemented with essential vitamins and minerals.

If you suspect laminitis, call your vet immediately and follow instructions. This will include taking the horse off grass straight away and stabling him on a deep bed. Depending on the severity of the case, treatment may include anti-inflammatory drugs and remedial farriery, so your vet and farrier will want to work together to restore the horse's soundness.

When you know a horse or pony is susceptible to laminitis, you need to manage him carefully. Preventive methods include:

- Don't graze him on land which has been sown with dairy cattle in mind. Grass that is lush enough to boost milk production is far too lush for horses.
- Restrict grazing by using electric fencing to strip graze and fitting a muzzle or mask that allows him to graze but limits his intake.
- If possible, stick to the safest grazing times when fructans are at their lowest, which will mean between overnight

between about 9 p.m. and 9 a.m. Don't turn out when there has been a frost, because as the grass warms up, fructan levels rise.

- Keep to a sensible, regular exercise regime- good advice in any case! Laminitis is often seen in ponies that have low workloads while their owners are at school and are then suddenly put into harder work during school holidays.

Mud Fever and Rain Scald

As you can guess from their names, both mud fever and rain scald occur in wet conditions and are skin infections caused by a bacterium which enters the skin through small abrasions. When the horse is exposed to prolonged rain, the skin softens and is more susceptible.

Mud fever develops on the legs and can range from mild irritation to open sores, swollen; painful legs; and lameness. Horses with white legs are particularly susceptible, as the underlying pink skin is more vulnerable than dark skin.

As soon as you spot the first signs of irritation—usually small sores which scab over—get your vet's advice. You will probably be advised to stable your horse for a while, if possible, but this isn't always practical. The general advice is to clip hair away from the affected area and as many horses will not tolerate clippers on a painful site it's often safer to do this with scissors which have curved blades.

Whatever method you use, be careful, as the kindest horse may lash out if caused discomfort. In some cases, you may have to ask your vet to sedate the horse so that clipping and initial cleaning of the area can be done safely.

Your vet will suggest a good antiseptic wash to remove the scabs. This is important, because unless the scabs are lifted, the bacterium will multiply. The skin should be dried carefully, either by blotting or fitting special leg wraps which wick moisture away from the skin to the outside of the fabric. Depending on the individual case, you may be advised to apply a nonstick dressing to the affected area, held in place by bandaging. Antibiotics may also be prescribed to help the horse hear.

Once the infection is under control, your vet may suggest topical preparations to act as a barrier. Some owners report good results from special turnout leg wraps used when horses are out to graze, but these must be the perfect size for the horse and adjusted correctly, because if dirt works its way underneath, abrasions may result.

Opinions vary on whether horses' legs should be washed off or left to dry naturally when they come in. From my experience, hosing off mud with cold water—not warm, which will open the pores of the skin—then patting dry or applying leg wraps works well. Don't use a brush with stiff bristles to remove mud, either when hosing or when the legs are dry, as these can cause tiny scratches and let the bacterium into the scratches.

Rain scald has the same characteristics as mud fever, but occurs on the face and body; the most usual sites are the back and hindquarters. Treatment is the same as for mud fever but prevention is much easier—bring susceptible horses in under cover during long or heavy periods of rain or use a waterproof blanket. There are now many designs of blankets for use in all conditions, from mild to cold, so you should be able to keep him comfortable as well as protect him.

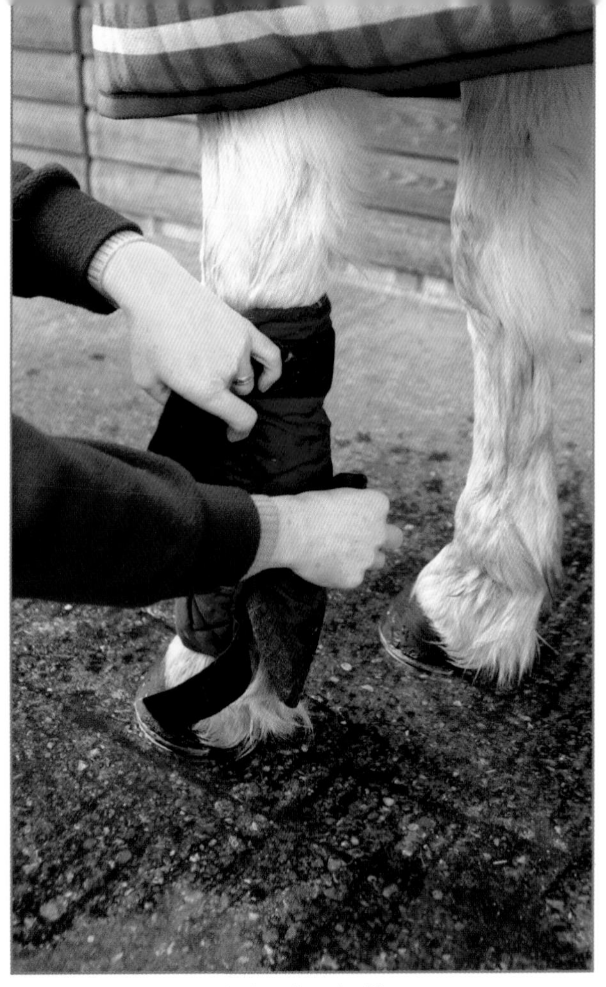

Special leg wraps can help dry off washed legs.

Respiratory Disease

Maintaining a healthy environment for your horse is an essential part of his daily care. Horses are naturally outdoor animals, but for convenience we sometimes stable them. This means their respiratory systems are susceptible to dust and spores from forage and bedding, so everything possible must to done to minimize these threats.

It has been estimated that up to 80 percent of horses and ponies suffer from some degree of Recurrent Airway Obstruction (RAO). In the mildest cases, their owners may not be aware of it but in more severe ones, the horse will show signs such as coughing and labored breathing.

If your horse shows signs of respiratory problems, get your vet's advice. Strategies may include keeping him out all or most of the time and, perhaps, drug treatment. However, every horse should be kept in a way that reduces the threats to his respiratory system. This means:

- Choosing bedding that is as free from dust as possible.
- Using a mucking out system that avoids the buildup of ammonia.
- Taking the horse out of the stable before you groom him.
- Feeding soaked hay or haylage.

Ringworm is a fungal infection of the skin.

Ringworm

Despite what its name implies, ringworm is nothing to do with a parasite, but is a fungal infection of the skin. While it is not a severe health problem in itself, it can cause a lot of problems—not only will it spread through a group of horses very quickly, it can be contracted by people and the fungal spores that cause it can survive for a long time on anything that the horse has come into contact with.

The first stage often appears as tufts of hair standing up from the skin. These tufts then fall out, leaving irregular-shaped patches of scaly skin. It can occur anywhere on the horse's body but is most often seen on areas where tack has been in contact, especially on the face and in the girth area.

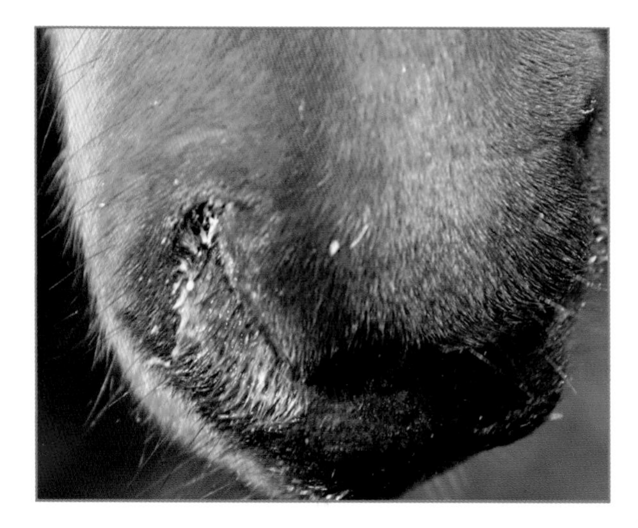

Although it will disappear over a period of about three months if left untreated, when treated immediately it limits the spread of the infection—it is unfair to knowingly let a horse with ringworm come into contact with others. Your vet will recommend a suitable treatment, which may mean topical application of a special anti-fungal disinfectant, drugs that are mixed in with feed, or both.

Anything that has come into contact with the horse—tack, blankets, grooming kit, etc.—should be treated with the anti-viricidal disinfectant and your vet will advise you on good hygiene practice. Ringworm is one of the best reasons for every horse having his own tack and equipment.

Strangles

A highly contagious respiratory disease, strangles is most often seen in young horses, though it can affect animals of all ages. The classic signs are swelling of the lymph nodes below the horse's throat and thick nasal discharge, but early indications include a raised temperature, loss of appetite, coughing, watery discharge from the nostrils, and general malaise.

It's important to get veterinary advice immediately if strangles is suspected and to isolate affected horses. Your vet will prescribe medication and advise you on how to maintain good hygiene practices to try and prevent the spread of the disease. Stringent hygiene is vital during a strangles outbreak, as bacteria are spread through direct contact and are released into the environment when the horse coughs.

If strangles is confirmed, it's also important to tell the owners of any horses which may have been in contact with the patient so that they can monitor them. Unfortunately, some people seem to think there is a stigma about the disease and try and keep it quiet. This is silly, unfair, and unreasonable, as strangles can affect any horse, anywhere, and is no reflection on the way an animal is looked after or where he is kept.

Sweet Itch

Every year, thousands of horses and ponies suffer from sweet itch, caused by an allergy to the saliva of the biting midge *Culicoides*. It causes mild to severe skin irritation, which prompts the horse to rub against fencing, trees, and anything else he can find and in the worst cases, sufferers will rub themselves raw.

Although all animals will be bitten by the midge, not everyone will react to the saliva and researchers now think that there may be a genetic element. If your horse is one of the unlucky ones, your vet will advise on management and in some cases, may prescribe medication such as antihistamines. It's important to take action as soon as you see the first signs and with care, you'll be able to keep your horse happy and itch-free.

Tactical management methods include:

- If possible, graze your horse away from natural water sources, as ditches and dykes provide ideal habitat for the midges.
- Keep water containers and troughs clean.
- The safest time to turn your horse out is mid-morning to mid-afternoon, as this will avoid the midges' favourite feeding periods.
- Invest in protective horse clothing designed to resist the midges and if possible, ensure that your horse wears it all the time he is not being ridden. You can find hoods, body wraps, and even leggings and udder covers for susceptible animals. Make sure they are made from fabric known to be resistant to the midges—ordinary anti-fly blankets, while an effective deterrent to some insects, may not provide enough protection against *Culicoides*.
- Use an effective fly repellent. Some owners report that adding garlic powder or granules to their horses' feed acts as a fly repellent, but opinions vary.

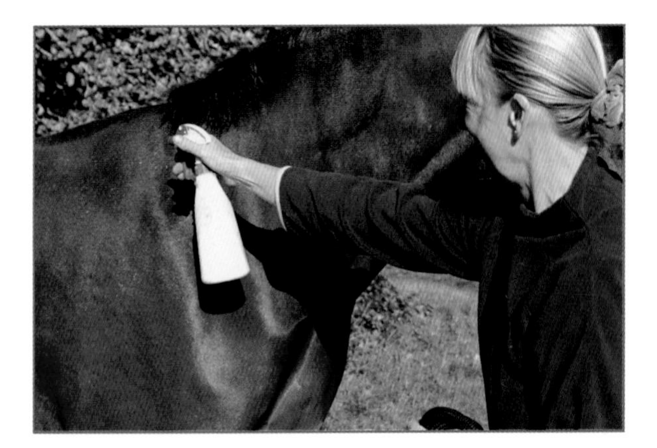

West Nile Virus

Over the past ten years, West Nile Virus has become endemic throughout the U.S. It is maintained by a life cycle involving birds and mosquitoes, and humans can also be affected. However, this can only happen if a person is bitten by an infected mosquito; it cannot be passed from horse to human.

Clinical signs include lack of coordination, weakness, muscle stiffness, and inability to swallow. In severe cases, the virus can cause inflammation of the brain or of the brain lining and spinal cord.

Disease prevention measures are vital and will contribute to your horse's overall health as well as protecting humans. Do everything possible to repel mosquitoes; in particular, reduce the number of potential mosquito breeding sites by keeping water containers clean and eliminating sources of stagnant water.

There are now vaccines in the U.S. which are reported to be successful and American vets and scientists are collaborating with those from other countries to build disease control strategies.

Wounds

Horses, like people, can be accident prone. No matter how well you look after them and how safe their environment, they will sometimes injure themselves—perhaps a horse will be kicked by a field companion, or tread on a sharp object too small to be spotted.

One of the most important things you can do for your horse is to make sure he is protected against tetanus. The cost of the vaccination is very low and could mean the difference between life and death.

Whether or not a wound needs veterinary attention depends on where it is sited as well as on its severity and if in doubt, you should always call your vet for advice. You can always email a picture of the wound!

In general, any wound that is deep enough to need stitching, has punctured the foot, or is near a joint or tendon means you should contact your vet immediately, even if it is small. If vulnerable structures have been penetrated, prompt action can mean the difference between treatment being successful and long-term complications. The outcome is always much better for wounds which are treated within the first few hours.

Minor nicks and scrapes can be treated by clipping the hair away from the site of the wound with a pair of round-ended scissors, then cleaning with dilute antiseptic. The antiseptic must be diluted at the correct rate, because a solution that is too strong can cause cell damage.

If a horse comes in from the field with a cut and his legs are muddy, first gently hose the area with clean, cold water or use a syringe to apply the water. Don't use a strong jet of water as this may drive dirt deeper into the wound. You can then clean with dilute antiseptic, making sure it is diluted at the correct rate.

Minor cuts and scrapes, once cleaned, often benefit from an application of wound gel, which allows the wound to heal in a moist environment. If you have any doubts about whether you can treat a wound yourself, or think it might need stitching, don't apply any gels, creams, or other products—clean the wound and call your vet.

If there is fast blood flow from a wound, the most important thing is to stop it while your vet is called out, as the blood will probably have cleaned most of the dirt out. This is particularly important with arterial bleeding, which spurts with each heartbeat. If you or someone else has been shown how to apply a pressure bandage, do so; if not, hold a clean cloth pad against the wound with enough pressure to staunch the flow and try not to release it until the vet arrives.

You may need to poultice a wound, particularly one in a foot. The easiest way to do this is with a commercial poultice dressing. A hot dressing to draw out infection is usually cut to size when dry, then soaked in a shallow container of hot water that has been allowed to cool to 100 degrees F (38 degrees C). It should then be removed from the container and the excess water squeezed out.

Apply petroleum jelly to the horse's heels to prevent them being rubbed or softened, then put the dressing in place with the plastic backing on the outside, away from the skin or sole of the foot. Some people then like to put a square of aluminium foil on top to hold in the heat as long as possible. Add a layer of padding, such as Gamgee, and bandage in place.

To keep a foot poultice clean, either use a proprietary poultice boot which fits over the foot, or wrap waterproof duct tape over the top. Unless your vet instructs you otherwise, change daily and stop poulticing after three days, as to continue doing so after this period means there is a risk of softening the sole.

If you find a horse is bleeding from his nose, seek veterinary advice immediately. Copious blood flow after fast exercise is usually due to exercise induced pulmonary hemorrhage (EIPH) but nasal bleeding can also be related to a blow on the head, such as a kick from another horse.

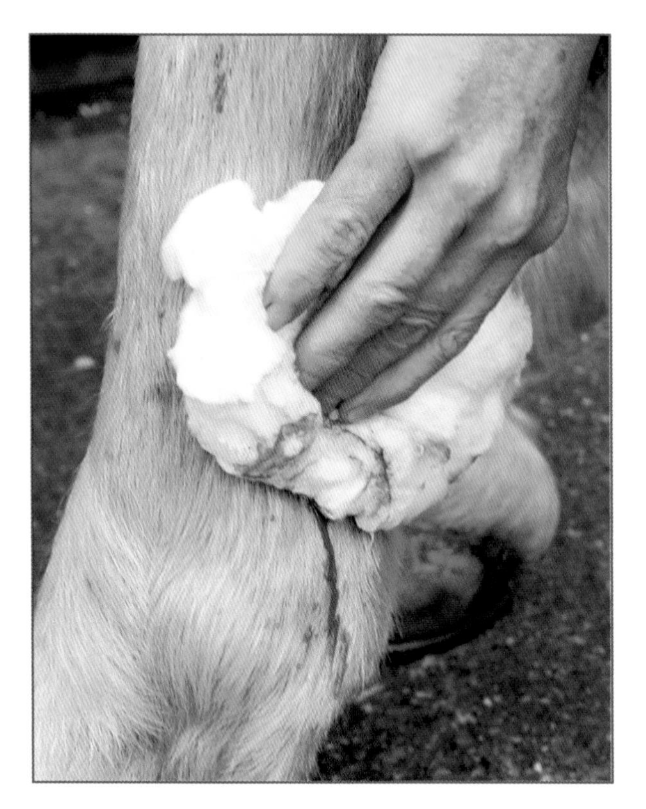

Complementary therapies

There is as much interest in and enthusiasm for complementary treatments and therapies for animals as there is for people and undoubtedly, they have much to offer. However, the difference between you and your horse is that you can communicate how you feel and a your horse can only show you by the way he reacts. It's therefore vital that if you think your horse has a problem, you start by getting a diagnosis and advice from your veterinary surgeon.

It may well be that a vet will recommend help from a qualified practitioner, such as a chartered physiotherapist, who will work under veterinary supervision. Some vets are also qualified in acupuncture, chiropractic, and so on and will use their knowledge and skills as part of a treatment regime when appropriate.

However, there are also people who claim to be able to improve a horse's well-being even though they have no recognized training or qualification. The golden rule must be—start with your vet!

Always stretch the horse's front legs gently forward after tightening the girth.

Hands-on Help

Simple massage techniques and stretches can help keep your horse relaxed and supple. Many competition horses have regular massages from qualified masseurs to help keep them in top working order and your vet or a practitioner working with your veterinary practice may show you how to carry out particular stretching techniques—in particular, those that involve moving a limb—as part of a general program or for rehabilitation purposes. Although there are many good books, videos, and DVDs that show such techniques, it's important not to go outside a particular range of movement and for this reason, a practical demonstration offers the safest way to learn.

Having said that, there is one gentle stretch that we should all do every time we ride. When you have girthed up your horse and are ready to mount, first stretch each of his front legs forward, slowly and gently. Pick up the leg as if you were going to clean out the hoof—being careful not to raise it higher than you need to—and gently stretch it forward to prevent the skin round the girth and elbow being pinched.

Keep your movements slow and steady and don't try and get more movement than the horse wants to offer. Hold for up to ten seconds, then gently bring the leg back and lower the foot to the floor.

Carrot stretches are useful techniques for suppling and stretching a horse and can be done before or after work, or both. Most vets and practitioners recommend these to all owners, as the horse can't be forced to work out of the range of movement he finds comfortable.

They get their name simply because they encourage the horse to stretch round or down to get a treat—and a supply of carrots is the easiest way to tempt him! Start by standing him on a level surface; to prevent him "cheating" and moving around to make things easier for himself, you may want to start in a stable.

The first stretch works by asking for side flexion through the neck. Simply use the carrot to encourage the horse to bring his head around to the side until his nose is at the point of his shoulder. You want to see him bending without tilting his head—if he really can't do it without head tilt, get him checked out by your vet in case there is a problem.

If all is well, repeat on the same side, then carry out the same stretch twice on the other side. Keep him interested by giving him a piece of carrot every now and then, but not every time.

Once he has the idea, and providing he seems happy, ask him to stretch around further so his nose reaches round towards his hip. It may take several sessions for him to

Start by asking for flexion to each side.

make the maximum stretch and he will have to tilt his head to achieve the range of motion. This exercise works on the whole back area.

Next, ask the horse to stretch his neck down by holding the carrot between his knees, positioning the treat so he keeps his head and neck straight. Once he understands what you can ask him to do, lower the carrot to ask him to stretch further. Not surprisingly, a lot of horses will want to bend their front legs to make it easier, but reward him only when he stands straight and doesn't bend a leg so he associates the correct stretch with getting a reward.

This stretch works the neck and back and encourages the horse to lift his back. To do this, he has to raise his abdominal muscles. Not only does this help his general well-being, it will help his athleticism when ridden. To work on the bit, in a rounded outline with his hind legs coming underneath him, the horse needs to lift his abdominals and raise his back, so exercising those muscles in the same way is helpful.

Massage

Massage is, as we all know, a way of relieving muscle tension and stress. Giving a horse a full body massage demands knowledge of anatomy and techniques, but there are simple ways to help your horse. One of the simplest is grooming: many horses enjoy the sensation of a rubber groomer being worked over their neck and body without weight being put behind it. This in itself is a light massage that can relax a horse.

Some horses enjoy being massaged with the flat of your hand, working in long, slow strokes. Start at the neck and stroke towards the tail, using a definite but not hard pressure. Experiment until you find the pressure your horse

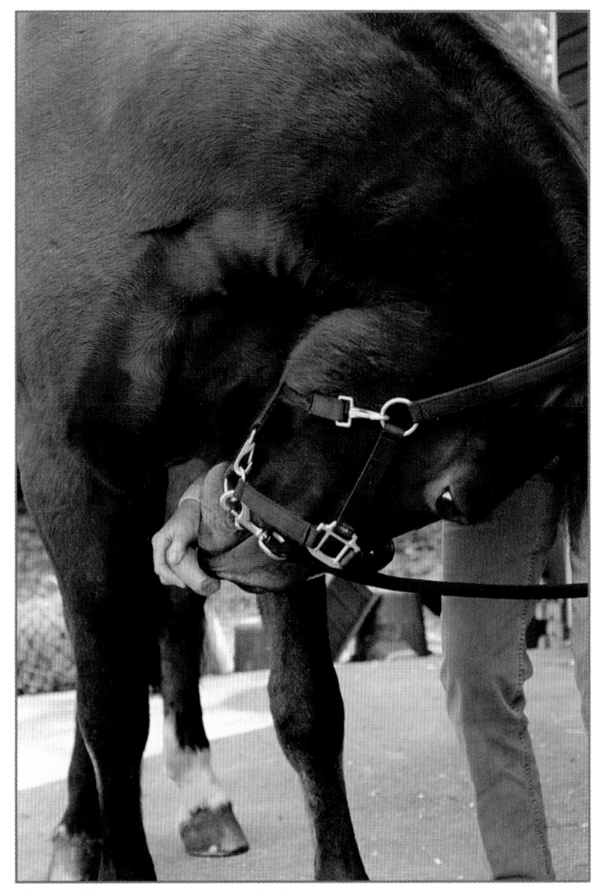
Next, ask your horse to stretch down.

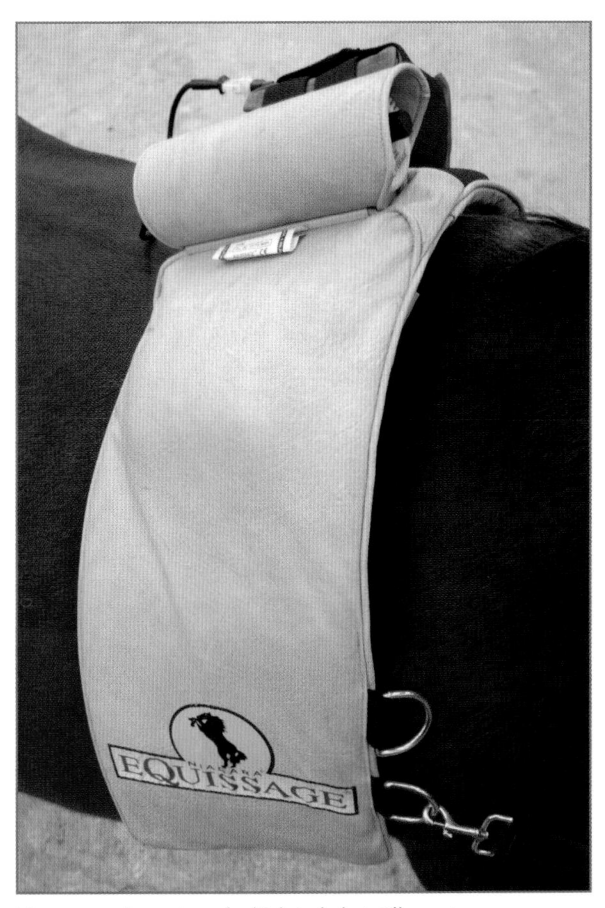

Massage equipment can be high tech, but still easy to use.

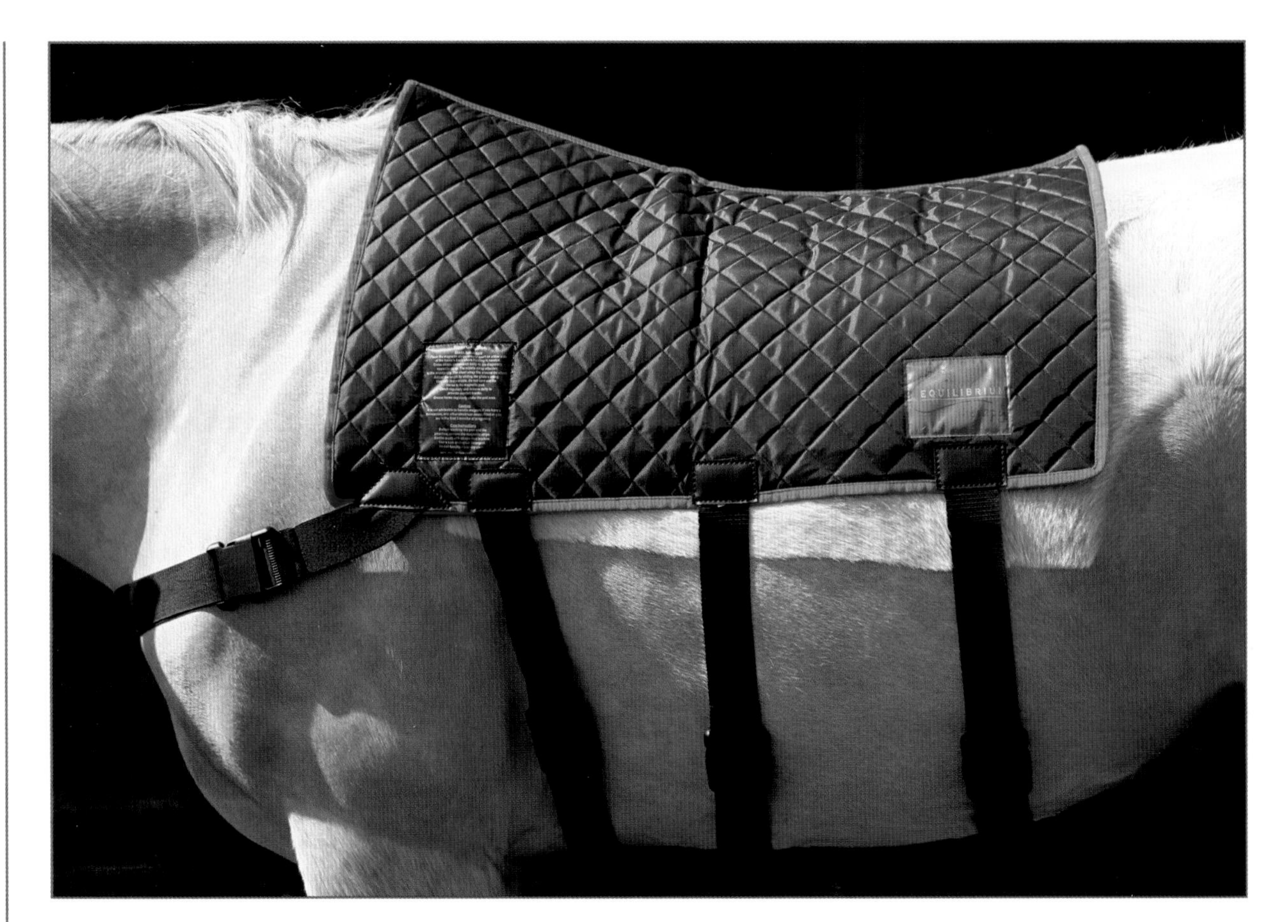

Magnotherapy is thought to increase circulation and in some cases, promote healing.

accepts best; you don't want him to flinch away from you, but nor do you want to irritate him by dabbing at him. When you've worked on the neck, go down the shoulders, then across and down the back and sides, and across and down the hindquarters. Work on each area on each side before moving on to the next.

Massage equipment has become part of many owners' regimes and can range from sophisticated and expensive setups to handheld units.

Magnetic Therapy

Equipment incorporating magnets—in particular, blankets, back pads, boots, and bridle headpieces—is used by many riders. The use of magnets to promote healing was recognized in ancient Chinese healing and it is now generally accepted that there is evidence that magnotherapy can help boost healing and well-being in some cases. As always, the first rule must be that before considering its use to try and help a problem, you should consult your vet.

In simple terms, magnetic therapy is said to improve circulation and thus have a healing effect, as the magnetic field improves the transfer of substances through the blood cells, speeding up the delivery of oxygen and nutrients and also helping the removal of toxins via the lymph system. Research using thermography (heat scanning) has shown that there can indeed be an increase in blood flow and scintigraphy (which uses radioactive markers) has demonstrated an increase in the rate that substances taken into the soft tissues are taken up by the body.

So when could magnetic therapy help? In general, if improved circulation would benefit a condition, it could be worth trying if your vet approves—so candidates may include horses suffering from tendon and ligament injuries and those who could benefit from increased blood flow before exercise, such as animals with stiffness and arthritic conditions.

Blankets incorporating magnets are said to encourage muscle relaxation and therefore may also have a calming effect. Many top competition riders like to use them before and after working their horses. However, because increased circulation means an increase in body temperature, it is not recommended that they are used when horses are traveling, as they may get too hot. There are also bridles with magnets incorporated into the headpiece. These are also said to have a calming effect, but opinions vary.

Some experts believe that there are instances when magnetic therapy should not be used, which makes it even more important to get veterinary advice first. For instance, it is usually recommended that it should not be used on pregnant mares or horses suffering from hematomas.

Fit for the job

To keep your horse sound and in good health, you need to make sure that he is fit enough for his level of work. This applies whether he competes regularly or goes hacking/trail riding three or four times a week.

A good description of a fit horse is one who can do the job expected of him without undue stress. At top level, getting a horse fit becomes more specialized—a three-day event horse would not necessarily be able to complete a 100-mile endurance ride just as a sprinter probably wouldn't be much good at shot putting! However, for riders who enjoy a mixture of activities and ride for pleasure rather than on a professional basis, a general basic fitness program will give you a good start.

A horse may well become fit enough for lower levels of competition and/or schooling work by doing his normal job, but it isn't fair to leave it to chance and if he has been off work for a long period, you will need to start from scratch. The time it takes to reach basic fitness, which should enable you to compete in riding club competitions or complete a 25-mile ride without your horse becoming too tired, will vary according to his age, previous workload, and to a certain extent, his type.

A young horse who has just been broke will take longer to reach basic fitness than one who has reached that point but been off work for a couple of months. Also, while it's perfectly possible to get cobs and other "coldblood" animals to a decent level of fitness, it will usually take longer than when working with a horse who has a high percentage of Thoroughbred or Arab blood.

You also have to remember that just because a horse is fit enough to cope with the physical demands of a job, that doesn't necessarily mean he will be good at it. For instance, he might be fit enough to perform a dressage test or go around a course of jumps, but unless he also has the necessary suppleness, strength, and balance, he won't be able to perform the job well.

You'll sometimes see suggested fitness programs based on weekly timetables. Although these may give a guideline, they don't take into account the fact that horses differ and so do their owners' lives. It's more realistic to think of building fitness through four stages.

Stage 1

The first stage of fitness comes through LSD—long, slow, distance work. Unless your horse can comfortably cope with an hour's active walking each day, this is where you need to start, ideally starting with half an hour and building up to an hour.

However, if your horse has had time off and hasn't read this book, he may decide that he would like life to be a little more exciting the first time you get on him. If you think this is likely to be the case, then for safety's sake, lunge him for a few minutes in an area with good fitting so he gets his brain in gear. In either case, it's sensible to fit protective boots so that if he has a buck and a kick or simply stumbles, he is less likely to injure himself.

If you're lucky enough to have access to a horse walker, this initial period can be made easier. Most horses, even naturally excitable ones, settle calmly in these and some seem to enjoy it. As with all work, make sure he spends an equal amount of time on each rein—if he walks for ten minutes in a clockwise direction, he should then do ten minutes counter-clockwise.

Most horses settle calmly to work on a horse walker.

Stage 2

The next stage means introducing short periods of trot, when riding out and in the arena or schooling field. Your trot should be active and rhythmical, but not hurried and if you live in an area with gradually sloping hills, trotting up these is a good way to encourage your horse to put his hindlegs underneath him and use his back end.

You can also introduce short periods of lungeing in walk and trot as part of your regime. Lungeing is deceptively hard work for a horse, so start with five minutes on each rein and build up gradually to ten minutes. If you're not experienced at lungeing, get someone to show you how to do it effectively and safely.

Working over poles on the ground, either on the lunge or in-hand, will add interest to your horse's work and build his athleticism.

Stage 3

If all is going well, you can now incorporate short canter sessions. This should be on good going, whether in an arena or outside and the canter should be as balanced and rhythmical as your

horse's stage of schooling allows. Don't do much cantering on the lunge, as this can put strain on the hock joints.

Whether or not you want to jump, you can now incorporate ridden work over poles. By this stage, you should be able, if you wish, to travel to a local competition and take part in a dressage test at an appropriate level for your horse's standard of education.

Stage 4

Now you can increase both the effort your horse has to put into his work and the length of time he is ridden for each day. Schooling sessions can become more demanding and can be combined with hacking, work over poles on the ground, and through jumping grids.

Even if you don't want to jump in competition, or feel reluctant to venture far off the ground, gridwork—where there is a gradual buildup to a line of fences at distances that make it easy for the horse to meet them correctly—will help your horse's impulsion and balance. Most really enjoy it, but it's important that gridwork is done under the supervision of an experienced trainer who knows how to adjust the distances to the length of the horse's stride.

To build your horse's cardiovascular fitness (heart and lungs) you can now start to lengthen your canter sessions. However, avoid the temptation to canter in the same places every time you ride out, or your horse will soon start to anticipate cantering.

Rider Fitness

While you're busy concentrating on your horse's health and fitness, don't neglect your own. Apart from the fact that you'll enjoy your riding more if you're reasonably fit and in shape, you'll also do him a favor. A fit, balanced rider is much easier for him to carry, while an unfit one who wobbles and thumps in the saddle can cause discomfort and muscle damage—some problems that are blamed on saddles are actually down to the rider, as explained in the next section.

For a start, is your own body in balance? Normal, everyday life can impose all sorts of stresses. For instance, if you sit at a desk all day, you might have got into the habit of rounding your shoulders and slumping, neither of which will do your horse any favors when you're in the saddle.

Similarly, if you favor one side of your body more than another, which most of us do, you'll put more weight on one stirrup than the other. Unless you recognize this and take steps to remedy it, you'll be causing problems for your horse.

This means it's a good idea for every rider to get an annual check from a qualified practitioner such as an osteopath who specializes in sports injuries. In some cases, you may need to be seen more frequently. If you can find a practitioner who rides, so much the better!

Once you've got your body in as good an alignment as possible, maintain it. One of the best ways to do this is to practize Pilates, a form of exercise which strengthens core muscles and improves balance and stability. Many international riders have discovered the benefits of this system, which has the added advantage of helping to prevent back problems, even in those who have already suffered them.

There are many DVDs which demonstrate Pilates exercises, but it is safer and much more beneficial to have lessons from a qualified teacher. A good teacher will ask you about any health problems you have suffered and show you how to carry out the exercises to the level you are capable of accomplishing.

Yoga can also be beneficial for riders, but only in terms of flexibility: it won't build core stability in the way that Pilates does. Again, it's best to have lessons from a qualified teacher.

The Alexander Technique, which helps to align the body, is practiced by many riders. If you're lucky, you'll be able to find a teacher who rides and can relate to your aims.

In terms of general fitness, a brisk 20 minute walk every day will improve cardiovascular fitness. So too will swimming and cycling. Swimming is particularly valuable, as it does not impose any stress on your joints. Cycling is better in this respect than running, but you need to find an exercise system that you enjoy and will maintain.

Want to revisit some childhood pleasures? Then try skipping and—wait for it—hula hooping! Skipping improves coordination and cardiovascular fitness, though like running, can impact your joints. Keeping a hula hoop in motion around your waist will improve balance and is also good for strengthening your core muscles.

Is your horse happy?

As well as giving your horse every chance of staying physically sound and in good condition, you need to think about his mental well-being. In many ways, the two go hand in hand, because complying with his basic needs will keep him relaxed and happy. Basically, he needs the chance to be a horse!

Sometimes, it's easy to get priorities mixed up, especially when you are keeping your horse on someone else's property and fitting in his care and exercise around your own family and work commitments. The following checklist will help you decide whether you are giving him a good lifestyle, or if you need to make changes.

Does he have sufficient turnout time in an appropriate environment? If limited grazing means you have to keep him stabled all or most of the time in winter, he's going to become stressed, irritable, and perhaps difficult to manage and ride.

Perhaps this is already a familiar scenario. If so, try and find a different home where he can live a more natural and happier lifestyle.

Is he in good—not fat—condition and fit enough for the work you are asking him to do?

Does he have a varied work regime? While schooling is important, most horses get jaded if asked to work in an arena every day and will be much happier if schooling is interspersed with hacking/trail riding. In any case, you can school your horse while you're out hacking—ask for a transition at a particular spot, practice leg yielding or shoulder-in in safe places, and so on.

If hacking is your main activity and you have a limited number of routes available, both you and your horse will enjoy traveling to different places to ride out. Depending on where you live, riding on a beach or through a forest can give you both a boost!

Stereotypical Behavior

Some horses become stressed in particular environments—usually when they are stabled—and demonstrate stereotypic behavior. There are three main behavior patterns: weaving, crib biting, and wind sucking.

A horse who weaves stands and repeatedly sways his head from side to side, usually over a stable door. In severe cases, he may shift his weight from one fore leg to the other.

Crib biting, where a horse seizes a ledge, door, or fence post, may be accompanied by wind sucking, where he gulps in air. Occasionally, a horse will wind suck without crib biting. Some horses will show these behaviors outdoors as well as in, but while they may crib bite or wind suck most of the time when stabled, will tend to do this less frequently when outdoors.

These patterns were commonly to be called stable vices and some people still refer to them in this way. However, that isn't fair to the horse, as they are only shown by domesticated horses and are rarely seen in ones which live outdoors all the time.

So why do they carry out what seems to us to be such bizarre behavior? It's now believed that the repetitive patterns stimulate the production of endorphins, natural chemicals that produce a "feel good factor" and that this is the horse's way of coping with a situation he finds stressful.

The traditional way of dealing with stereotypical behavior was to physically prevent the horse from doing it. A V-shaped, anti-weaving grille on top of the stable door will discourage him from swaying his head from side to side, though some horses will simply stand back and weave behind it. It can't be classed as cruel and may deter a mild weaver, but it won't decrease his stress levels and is probably of more benefit to an owner or manager who doesn't like to see the horse behaving in this way.

When it's safe to do so, a stallguard placed across an open doorway often reduces a horse's compulsion to weave. Special stable mirrors made from reinforced materials have also been shown to cut down weaving and crib biting in some horses. It is believed that the horse takes comfort from his reflection, believing it is another horse. Unfortunately, mirrors can't always be used with stallions, as some behave aggressively towards their reflections.

Where possible, installing a grille between compatible horses in neighboring stables can help, as being able to see each other and "talk" through the bars reduces anxiety levels. However, if two horses take a dislike to each other or one is jealous of his food, this can cause more problems than it solves.

For many years, anti-cribbing collars were recommended as a way of dealing with the problem, but the traditional design is not recommended on humane grounds. This comprises a broad strap with a metal arch which is strapped round the horse's neck so the arch sits round his throat. When the horse arches his neck to take hold of the door or whatever else he wants to crib bite on, the metal digs into his throat. The discomfort might discourage him, but it won't do anything for his stress levels!

The best way of preventing or at least decreasing stereotypical behavior is to turn the horse out as much as possible; when practical, a 24/7 outdoor lifestyle in a well-managed environment is the best one for a weaver, as few horses weave outdoors. Crib biting is not so easy to deal with, as some horses will crib on fences and gate posts; however, most do so only sporadically and again, it is recommended that they are turned out as much as possible.

Painting foul-tasting but harmless anti-cribbing products on stable doors may dissuade the horse, but he will only look for another site. While a horse will obviously not crib bite on electric fencing, he may crib on the posts used to support it.

Crib biting alone will cause wear on a horse's teeth, but this in itself does not usually cause problems. Wind sucking, where the horse gulps down air, is sometimes related to cases of colic.

Some people worry that horses copy stereotypical behavior from others, but it has been proved that strictly speaking, this is not the case. Researchers now believe that a horse will only start this behavior if he already had a propensity towards it.

All in all, keeping a horse happy is a vital part of good management. Just remember, when you're choosing a home for him and working out the regime you intend to follow, that you apply equine values rather than human ones.

The hardest decision

Sadly, at some point in your horse owning life you may have to make the hardest decision of all and acknowledge that it is time for your horse to be put down.

In some cases, this happens through injury or illness, in which case your vet will explain to you why it is the only option. In others, it will be a decision you have to make, though your vet will help you come to it. Horses rarely die from natural causes and usually you will get to the stage where an animal no longer has a quality of life, perhaps because he is in constant pain from arthritis, because his eyesight is failing, or simply because he is no longer able to keep adequate weight on even when fed and cared for correctly.

Many horses and ponies will live happily in retirement for some years after they are no longer able to work and if you are able to give your horse a quality retirement, that may well be the option you choose. You can't simply leave a horse out in a field and expect him to be happy. He will still need routine preventive care such as worming, vaccination, and dentistry and he will still need his feet trimming. In cold, wet weather he will need shelter and blankets and in summer, he will need protection from flies and biting insects. Occasionally, horses who have led active lives sometimes do not adapt to retirement and are obviously unhappy.

Having a horse put down may be the last thing you can do for him, so please don't walk away from the responsibility. It is unfair and cowardly to try and convince yourself that he will be able to do light work with someone else if you know this is not the case. There have also been cases of unscrupulous people offering to take horses as "companions" and selling them, often at auctions, as riding animals.

There are two methods of euthanasia: shooting or by lethal injection. Shooting can be carried out by a vet or licensed slaughterer and death is instantaneous, though the noise, sudden collapse of the horse, and reflex movement after death makes it unpleasant to witness. The injection method takes longer and occasionally, a standing horse may stagger before he falls.

Although it isn't a subject anyone wants to think about, everyone who owns a horse needs to do so. Try and discuss the options with your vet when he or she makes a routine visit in case it is something you need to decide on—hopefully, a long way down the line. A good vet will be pleased that you are being responsible and will talk you through the options, including available methods for disposal of the horse's body.

Some owners may decide they want to be with their horse when he is put down. In most cases, it is better to say goodbye and leave him in the care of your vet or the slaughter, who will be compassionate but will remain calm. Horses pick up on distress signals and it is kinder to your horse to keep out of the way. Remember that he doesn't know what is about to happen and will not be frightened.

You may have to decide whether an old horse still has a quality of life.

PART 5

TACK AND EQUIPMENT

It will soon become obvious that looking after and riding a horse calls for a lot of equipment and that it's very easy to spend a lot of money very quickly. However, although every horse owner gradually collects more items, it's important to work out the essentials and to make wise purchases.

Just as important, you need to know how to choose the right equipment for your particular horse, how to fit it, and how to look after it. This section will help and you should find that a good saddler will be able to advise you, whatever budget you're working with. If you buy a horse, you may be offered the chance to buy his tack and other equipment with him and if it fits well and is in good condition, this may save you some money.

Good quality tack that is well-cared for lasts many years and in the long run, will be a better, safer investment than that of inferior quality. The most expensive individual item you'll need is a saddle, so if you're on a tight budget, ask your saddler about buying a secondhand leather or a new synthetic saddle. Avoid the temptation of real and internet auctions for the moment—you might get a bargain but you might get faulty goods. Buying from a specialist retailer should mean that a saddle has been checked over and necessary repairs carried out.

Modern synthetic designs, pictured above, are smart and easy to look after and although a synthetic saddle might not last as long as a good leather one, it will give good service.

In the hot seat

There are three golden rules about choosing a saddle: it must fit the horse, fit the rider, and be suitable for the work the horse is doing. For the first-time owner, a general purpose (GP) saddle such as the synthetic one pictured is the answer; as its name suggests, it is suitable for all-round riding and can be used for hacking, schooling, and jumping.

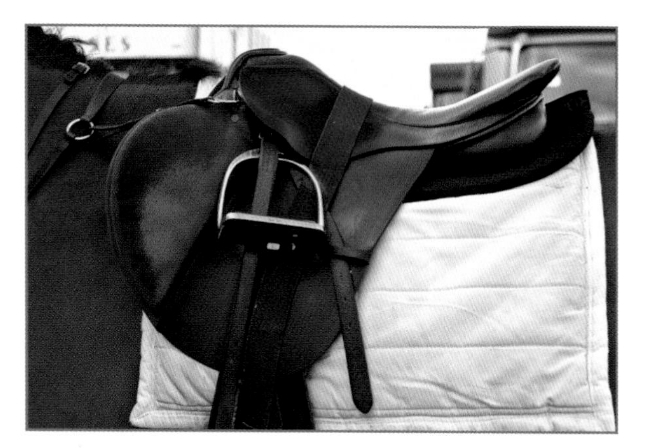

Experienced riders who specialize in dressage or jumping use specialist saddles. Dressage saddles have straighter flaps to encourage a correct position when riding on the flat; the rider will want to be upright, balanced, and have a longer, straighter leg position. Jumping saddles have forward cut flaps, as the rider will

have shorter stirrup leathers and the angles of the knee and thigh will be more closed.

Your saddle must fit your horse correctly and it is vital that you get it fitted by a professional saddle fitter. It is a big advantage if the saddle fitter also has a good knowledge of riding. This means he or she can check that a particular make of saddle encourages you to adopt a correct position and gives you enough security.

At the same time, you need to understand the basics of saddle fitting, as explained in the guidelines below, so you can recognize when you might have a problem or when your saddle needs adjusting.

Don't underestimate the problems a badly fitting saddle can cause. Even mild discomfort will probably make a horse resist what you ask him to do—and if fitting problems are not recognized, they can eventually cause long-term and even irreparable damage to muscles and ligaments.

All saddles need regular checks and adjustments. For instance, a synthetic saddle might be lightweight, but it can still cause pressure points if it does not fit the horse.

You will need to have your saddle checked regularly, as horses change shape as they gain or lose weight or muscle. Your saddler will advise you on how often your horse needs to be seen, though if you notice any problems you should get help straight away. Usually, a simple adjustment will sort things out.

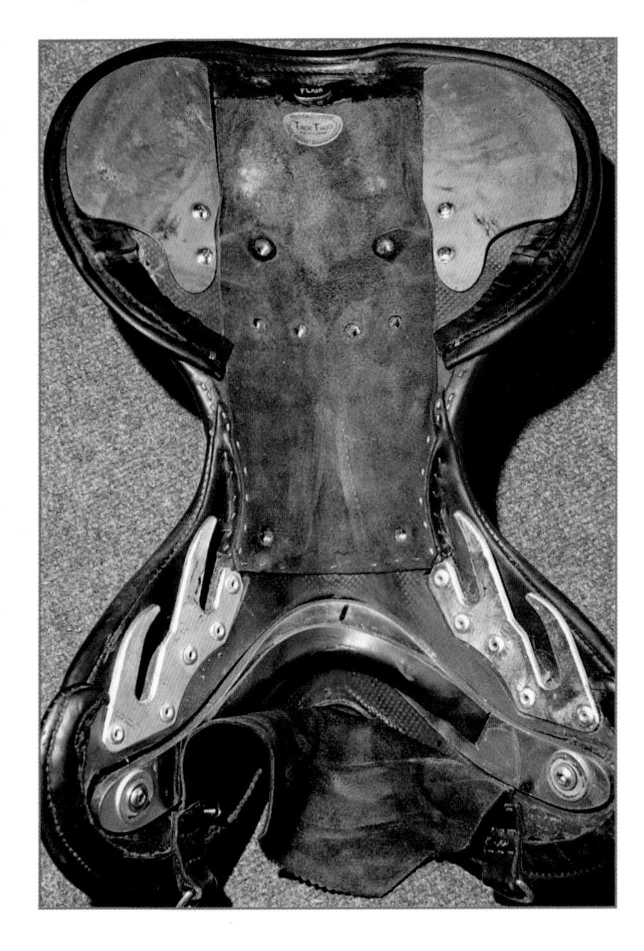

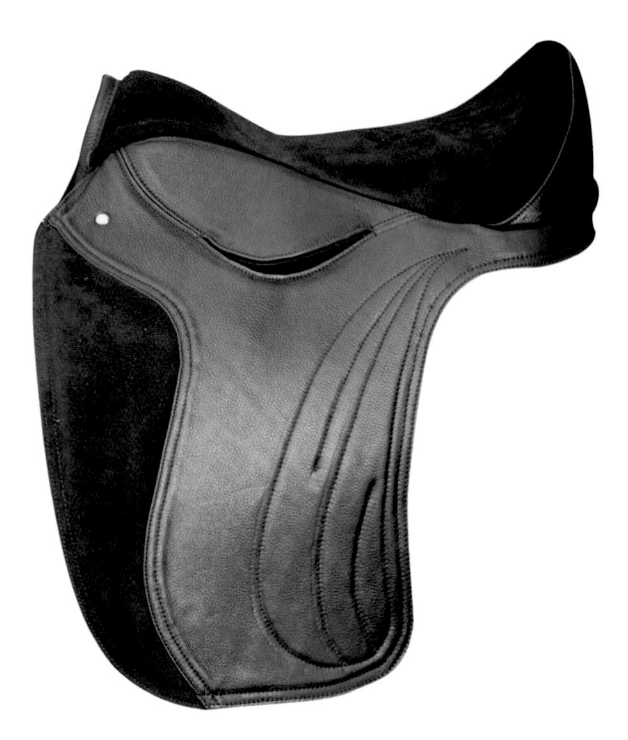

Most saddles are built on a frame called a tree. Trees are made in different widths and it is important that this is correct for the individual horse. If it is too narrow, it will pinch and if it is too wide, it will come down too low, again causing discomfort.

The tree must also follow the shape of the horse's back. If it has a curved profile, the saddle will not fit a horse with a flat back, and vice versa.

You can also buy saddles that are made without trees, or with partial trees. Although some people believe that they will fit any horse, they must still be adjusted with care and may need to be used with special pads to ensure they are correctly balanced and do not cause pressure points.

In general, a saddle should:

- Spread the rider's weight over as wide an area as possible.
- Not restrict the horse's movement.
- Be level from front to back, with the center of the seat the lowest point. If it is out of balance, the rider will be tipped forward or backward.
- Sit evenly, not over to one side. However, a rider who puts more weight in one stirrup than the other can cause this problem.
- Clear the horse's withers and back sufficiently so that it never presses down on them.
- Stay stable. There will always be some movement when you ride, but the saddle should not rock from side to side or bounce up and down at the back.

Although the horse's comfort must be given priority, a saddle must also fit the rider. This will also help the horse, because a rider who is uncomfortable will not sit in balance.

Key points are that the saddle should have a large enough seat to accommodate the rider comfortably—though a too large one will mean lack of security—and the proportions of the flaps should suit those of your legs. It's

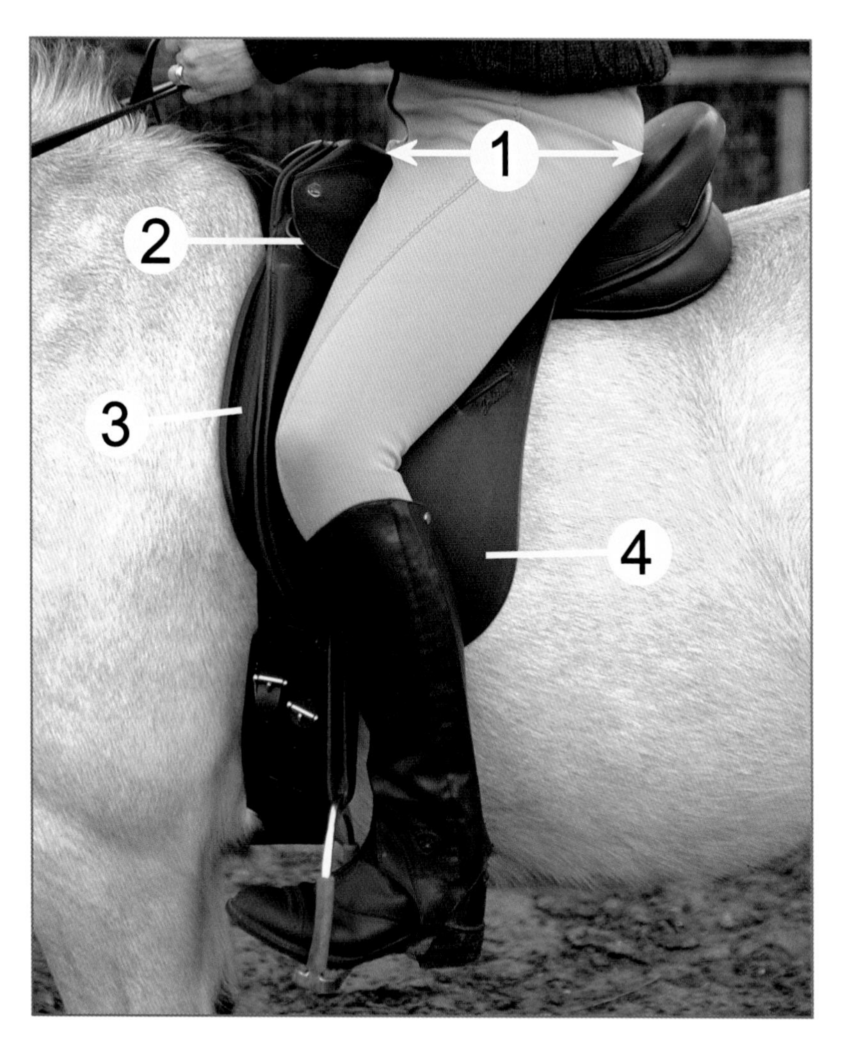

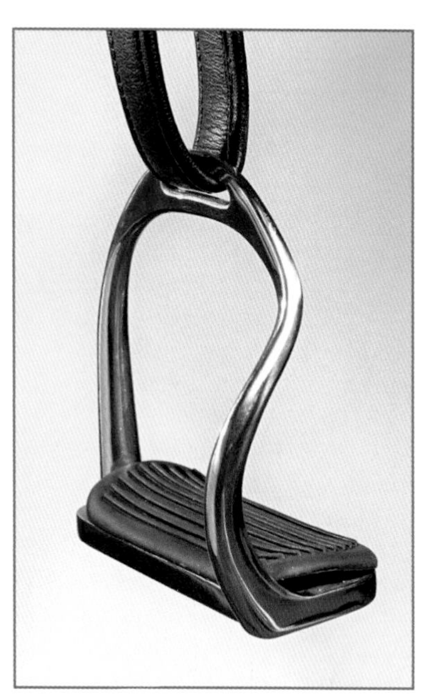

Left:
1 Seat size correct and puts rider in center of saddle.
2 Stirrup bars positioned to allow correct leg position.
3 Knee blocks positioned to allow legs to sit just behind them.
4 Flap proportions suitable for those of rider's legs.

also important that the stirrup bars, from which the stirrup leathers and irons are suspended, should be positioned correctly on the horse.

If they are too far forwards, you will tend to sit too far back and push your leg too far back. If they are too far back, you will probably tip forwards and when you're jumping, your lower leg will tend to slip back, interfering with your ability to balance. Your saddle should encourage a correct riding position, not hinder it, which is why you should look for a good saddle fitter with a knowledge of riding. If necessary, ask your instructor to assess your riding when trying different designs.

Saddle Extras

You'll also need a girth to hold the saddle in place, stirrup irons and leathers, and probably a saddle or pad of some kind. Always go for good quality accessories, as your safety depends on them.

There are several designs of stirrup iron, but whatever you choose, make sure the irons are the right size for your foot. There should be half an inch (1.25cm) clearance on either side of the widest part of your riding boot sole, but they should not be any wider than this or your foot may slip too far forward.

Even though correctly sized irons should minimize the risk of your foot getting trapped in the event of a fall, it's sensible to choose stirrup irons with extra safety features. These include bent leg safety irons and designs with hinged sides that click open if sufficient pressure is put on them. Safety irons which have a rubber ring on the outside should only be used by small children, as this design cannot take the stress of heavier weights. Rubber stirrup treads which slot into the irons add comfort and security.

Stirrup leathers should be an appropriate length for the rider's height. If you buy a synthetic saddle, you may find that there are synthetic stirrup leathers to go with it. However, don't use these on a leather saddle, as some materials may damage leather.

Girths can be made from synthetic

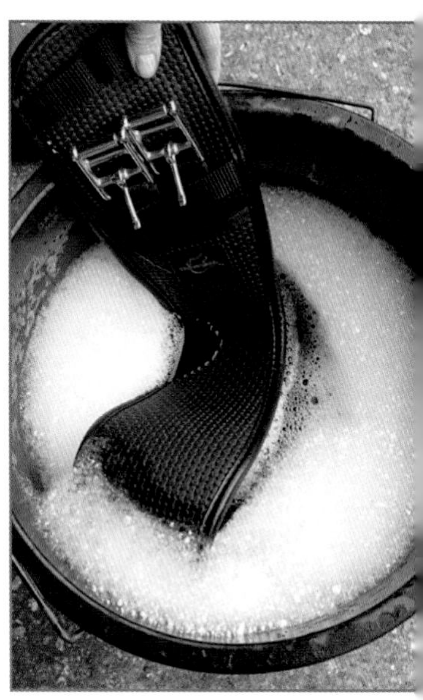

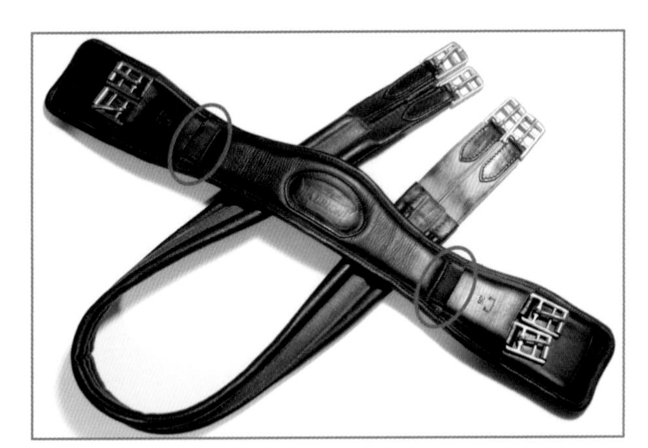

A girth with give throughout its length is more comfortable for the horse.

materials or leather. For everyday use, a soft, washable synthetic girth is easy to look after and comfortable for the horse. It's generally accepted that a girth with "give" throughout its length is more comfortable for the horse, but avoid ones with elastic ends on one side only, as they may encourage the saddle to slip to one side.

Most riders like to use a saddle pad to help absorb sweat and keep the underside of the saddle clean, but there are a few pitfalls to avoid. Don't use a very thick one without your saddle fitter's approval, or you could alter the fit of your saddle—rather like wearing two pairs of socks inside shoes that fit perfectly with just one pair! Also, make sure it pulls up into the saddle gullet and stays there; many designs slip down and put pressure on the withers.

Saddling Up

Take time when tacking up to make sure your saddle is correctly positioned and adjusted. Remember to start by placing it—with the saddle pad, if used, held in position underneath—forward of its final position, then slide it back so the horse's coat lies flat underneath. There should be a hand's width between the horse's scapula (shoulder blade) and the saddle, to allow room for the shoulder blade to rotate back as he moves.

Tighten the girth gradually. This will be more comfortable for him and he may be less inclined to blow out his belly; if necessary, spend five minutes girthing up and walk him around before adjusting the girth a hole at a time. Don't overtighten it—it should keep the saddle secure, but not cut the horse in two!

You may find that no matter how skilful your saddle fitter, your saddle tends to slip slightly forward or back because of your horse's conformation. If it slips forward, try attaching the girth to the front two straps on each side and if it slips back, use the rear two straps. This assumes that your saddle has a standard arrangement of three girth straps; there are other systems designed to give more scope for adjustment, as your saddle fitter will explain.

Whenever possible, use a mounting block rather than mount from the ground, as this puts less strain on both the saddle and the horse's back. If you have to get on from the ground, spring up, keep your weight forward as you put your leg over the horse's back, and don't pull yourself up by the cantle.

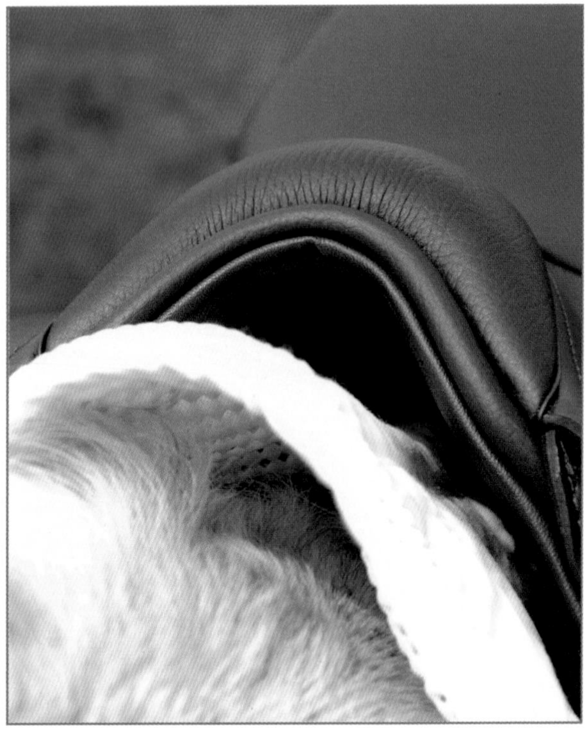

A saddle pad should stay in place and not put pressure on the withers.

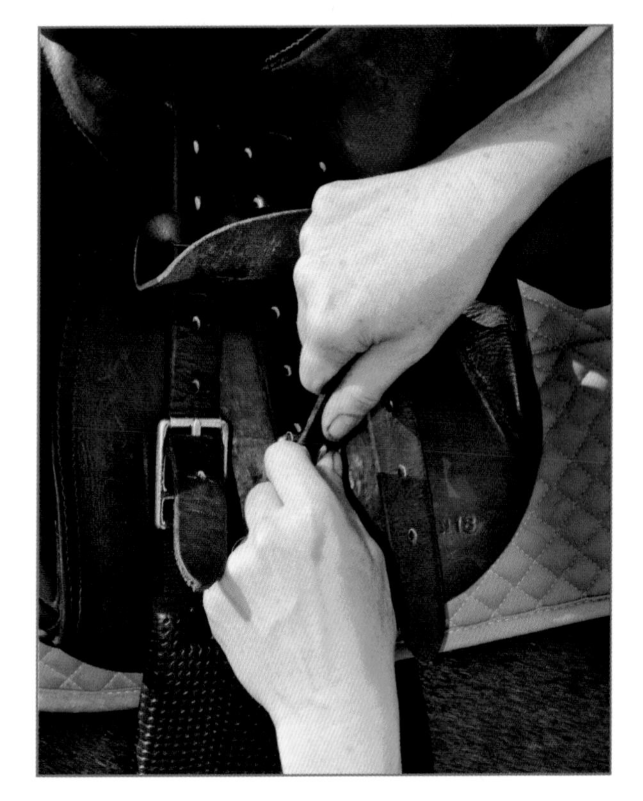

Bridles

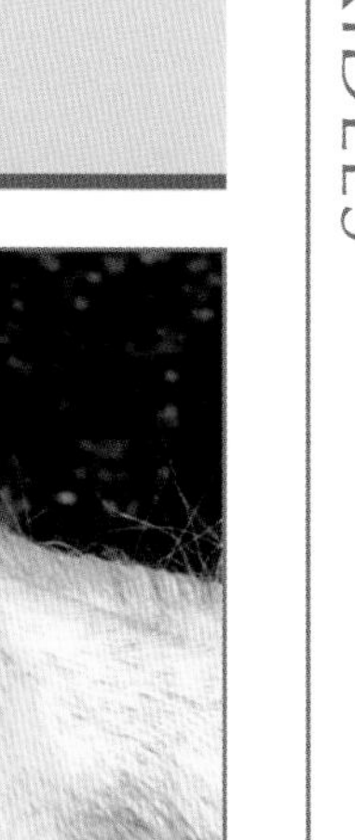

Whether you use a top quality leather bridle or a cheaper synthetic one, it's vital that it fits correctly. Parts of a bridle and the key fitting points are pictured here, but to get a perfect adjustment, you may need to mix and match parts from different sized bridles. For instance, a pony or cob with a broad forehead may need a browband from the next size up and a Thoroughbred with a narrow muzzle may need a full size headpiece, browband, and cheekpieces but a cob size noseband.

There are several common fitting mistakes, all of which may affect your horse's comfort and his willingness to work correctly. From the top, a browband that is too short will pull the headpiece on to the ears. If his ears are pinched, you can't blame a horse for throwing his head about and being generally resistant.

Many riders fasten the throatlatch (which for some reason is pronounced throatlash) too tight. There should be four fingers' width between the throatlatch and the side of the horse's face, to ensure that he can flex his jaw.

Nosebands should be high enough to prevent the bottom of the facial bones being rubbed, but not so low they interfere with the action of the bit or even the horse's breathing. Nor should they be too tight.

The simplest design of noseband is the cavesson, a plain strap which fastens above the bit and is used more for the sake of appearance, to give a finished look to the horse's head than for practical reasons—though as explained later, it can be used as an attachment point for a standing martingale, which, like other types of martingale, helps prevent the horse raising his head above the angle of control.

Other nosebands fasten below the bit. Their purpose is to give the rider more control by preventing the horse from opening his mouth too wide. It's important to remember this when adjusting one and not think in terms of strapping the horse's mouth closed. A noseband that is too tight will prevent him from flexing his jaw, which will also prevent him from staying relaxed and responsive to signals via the bit.

Above:

1 Browband long enough to prevent ears from being pinched.
2 Throatlatch fastened to allow a hand's width between it and the horse's face.
3 Top part of noseband must not rub facial bones or interfere with action of the bit.
4 Bottom strap discourages horse from opening his mouth too wide, without being tight.

The most commonly used noseband of this kind is the Flash, which is a cavesson noseband with an additional strap which fastens below the bit. Choose one with a substantial cavesson part, as a flimsy noseband will slip down the horse's face. The Flash noseband was originally

A standing martingale could be fastened to the top part of a Flash noseband.

designed for a show jumper of that name, so that a standing martingale could be attached to the cavesson part.

The drop noseband is a classic design that, when correctly made and fitted, can help encourage the horse to adopt a lower head position. When he opens his mouth, the noseband acts low down on the front of the nose and also in the curb groove at the back of the jaw, where there is an acupressure point.

A drop noseband must be fitted so that the top part rests on the facial bones. If it is too low and sits on the fleshy part underneath, it will interfere with the horse's breathing. For the same reason, never attach a standing martingale to a drop noseband.

The third design in common use is the Grakle, originally designed for a hard pulling racehorse. It's also referred to as a crossover or figure-of-eight noseband and is intended to dissuade a horse from crossing his jaw to evade the rider's aids. The top strap of the original design is fastened just under the facial bones, but some horses respond better to a high ring or Mexican Grakle—perhaps because there is no risk of pressure on the cheek teeth.

There are many more types of nosebands, but don't be in too much of a hurry to venture beyond the plain and simple. If a horse opens his mouth to resist rein pressure, first check your riding: if your rein contact is too strong or uneven, changing your horse's noseband won't make things any better for him.

A drop noseband must not impinge on the horse's nostrils.

A Mexican Grakle does not put pressure on the cheek teeth.

Choosing and Using Bits

Bitting is a subject that strikes terror into the heart of many experienced riders, let alone more novice ones. It's hardly surprising, because if you walk into any tack shop, you'll see a huge array of designs and seemingly endless choices of mouthpieces, materials, and cheekpieces.

There are four key points that will help you keep your horse comfortable and get the best communication with him. The first is that his mouth and teeth must be in good condition, which regular checks by a good equine vet with an interest in dentistry or a qualified equine dental technician will take care of. The second is that the design of the bit should suit the shape of your horse's mouth and the third is that it must be the correct size and adjusted at the right height.

The fourth thing to remember is that a bit can only be as effective as the rider using it. This means making sure you can ride in balance without relying on your reins and that you understand how to use your weight and leg aids to influence your horse; rein aids should be used with finesse, not to pull the horse in the direction in which you want him to go.

It's also important to remember that horses have to be taught what a rider's aids mean and that those who have been schooled to understand and accept subtle signals won't appreciate a heavy-handed rider. Those suitable for relatively novice riders should be fairly tolerant, but a young horse must be educated by an experienced trainer.

You must be able to ride in balance, without relying on your reins for support.

Fitting a Bit

The length of a bit is measured as shown and its thickness is the diameter of the widest part of the mouthpiece, near the rings or cheeks. Curb and shanks (pelham cheeks) are usually measured from top to bottom.

The mouthpiece of a bit should be long enough to prevent pinching, but not so long that there is excessive side to side movement. A useful guideline is that when a loose ring jointed bit is straightened in the horse's mouth, there should be no more than half an inch (1cm) between the holes on each side through which the rings pass and the horse's lips. A bit with fixed cheeks, such as the one shown here, can fit slightly more snugly but should not pinch.

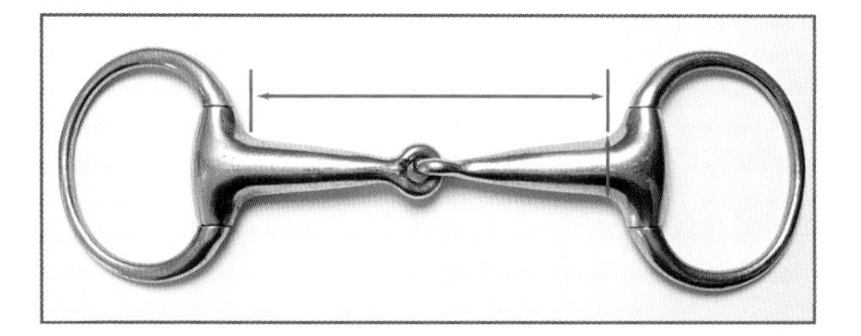

When adjusted at the correct height, a bit will lie comfortably across the bars of the mouth. If it is too low, he will be able to put his tongue over it and if it is too high, it will pull his mouth into a false smile. However, a horse with a lot of flesh round his lips will show more wrinkling round the bit than one who has only a little.

The Bit Families

There are literally hundreds of types of bits—and just when you think no one can possibly think of a new design, one appears on the market. To simplify things, it's easiest to think of bits as being grouped into four main families: simple snaffles, pelhams and kimblewicks, double bridles, and gag bits. You may also come across lever snaffles, combination bits incorporating special nosebands, and bitless bridles.

The snaffle is the simplest form of bit and is used with a single pair of reins. At the other end of the scale, the double bridle offers the most finesse in communication between a skilled rider and a well-schooled horse. It comprises two bits, a thin snaffle, which in this context is called a bradoon, and a curb bit, which has a lever action.

The pelham is an attempt to combine the actions of a double bridle in one mouthpiece and the kimblewick is a variation on it. Gag and lever snaffles and combination bits are designed to give extra control over strong horses and vary in their actions, as explained later in this section.

Bits can act on several control points, including the bars of the mouth, the lips, the tongue, and the curb groove (at the back of the jaw). Before deciding what type of bit to use, look at and in your horse's mouth—but be gentle and don't put your fingers in his mouth, because he could bite without meaning to. To get an understanding of mouth conformation, you need to look at different types and breeds of horses to appreciate the differences.

Does he have a lot of flesh round his lips, as is often the case with cob and draft type horses, or is it more taut, as with many Thoroughbreds? You need to be aware of this when adjusting the height of a bit.

Does his tongue sit comfortably in his jaw, or does it bulge out at either side? If he has a fat tongue, he will usually be more comfortable with a thinner mouthpiece rather than a thicker one. If you use a snaffle, he will also probably respond better to one that is double-jointed rather than single-jointed.

Happy Mouth pelham

Kimblewick

Lozenge snaffle

D-ring flexible snaffle

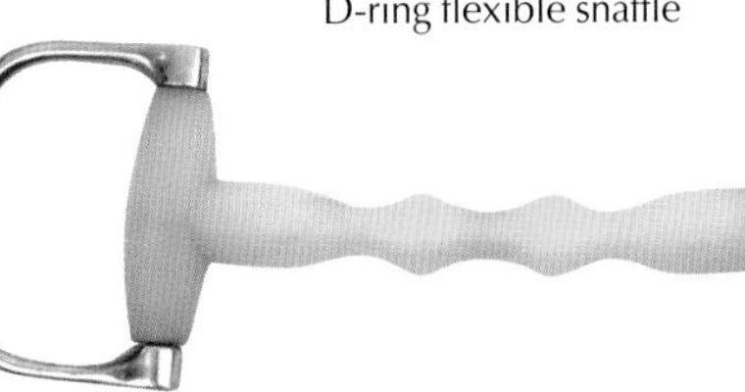

Mouthpieces, Cheekpieces, and Materials

The design of the mouthpiece and cheekpiece and the material from which it is made affects the way a bit works. It's often said that a thick mouthpiece is kinder than a thin one, because it has a greater bearing surface. However, as explained earlier, a horse will only be comfortable with a thick mouthpiece if he can accommodate it. If he can't close his mouth comfortably around the bit, a thinner one will be more appropriate.

A single-jointed bit acts on the corners of the mouth, the tongue, and the bars. A double-jointed snaffle, such as a French link or lozenge snaffle, reduces tongue pressure and many horses go well in it.

Some bits have mullen (slightly arched) mouthpieces, which allow room for the tongue and give a more direct signal than that offered by a jointed mouthpiece when the rein pressure is applied. Others have ports, which is a small arch in the centre of the mouthpiece that allows room for the tongue.

One design theory that seems to make sense is that a bit should follow the contours of a horse's mouth. This happens to a certain extent with unjointed, flexible mouthpieces, which bend slightly when the reins are used. It is also a feature of Myler bits, which come in a range of mouthpieces—the one shown here has a port.

Although the rider's hands will be the deciding factor, cheekpieces affect the degree of bit movement in the mouth. For instance, a loose ring allows for constant slight movement and would often be preferable for a horse who is "dead" in the mouth and/or tries to lean on the bit. A fixed cheek, such as an eggbutt or full cheek, means the bit remains more stable and a full cheek can also help with steering by putting slight pressure on the side of the mouth and muzzle.

Finally, think about the material the bit is made from. Traditional stainless steel is still a popular choice, but many riders (and horses) prefer materials which encourage salivation. These include copper and sweet iron.

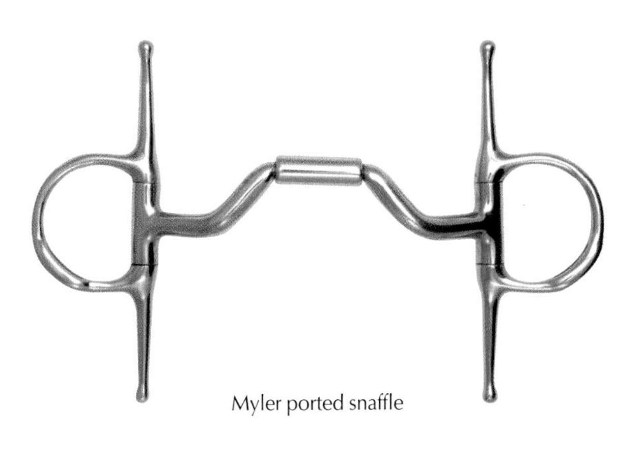

Myler ported snaffle

Loosering, sweet iron snaffle

Why do you need a horse to salivate? Quite simply, because saliva is a lubricant and a bit will slide over the bars of a wet mouth, but drag on a dry one.

Simple Snaffles

For most horses and riders, some type of simple snaffle is the best choice. Some traditional and newer ideas include:

Eggbutt Snaffle: This stays still in the mouth and the smooth eggbutt cheeks minimize the risk of the lips being pinched. For this reason, it is a good bit for novice and other riders whose balance is not established. It may also give confidence to a horse who "backs off" the bit.

Loose Ring Snaffle: The constant movement of this bit encourages salivation and discourages a horse from leaning on or setting himself above the bit.

Full Cheek Snaffle with Keepers: Leather keepers attach the upper ends of the bit cheeks to the bridle cheekpieces. This helps to keep the bit at the correct height in the mouth, thus discouraging the horse who tries to put his tongue over the bit and also decreases tongue pressure slightly.

Full Cheek Snaffle Without Keepers: The full cheeks keep the bit central in the mouth and help to reinforce steering aids.

Reinforced Plastic and Polymer Snaffles: These often make good introductory bits for young horses and for those with sensitive mouths. Their drawback is that they are not resistant to horses' teeth and can become damaged, so they must be discarded if this happens. For safety's sake, only use bits which have a metal core to prevent them being chewed right through when in use.

Hanging Cheek Snaffles: The design of the cheekpiece, rather like the top part of a pelham (see below) means that the bit is suspended in the mouth and lessens the amount of tongue pressure. However, when the reins are used, pressure is applied to the horse's poll through the bridle headpiece, which should encourage him to lower his head. Some horses respond well to poll pressure but others dislike and resist it.

Gag and Lever Snaffles

A true gag snaffle has slots in the cheeks to take running bridle cheekpieces. These may be made from rolled leather or cord and attach to a pair of reins. When the reins are activated, the bit slides upwards in the horse's mouth and poll pressure is also applied. The action of this bit is contradictory, because it employs raising and lowering actions at the same time, but it is sometimes effective on hard-pulling horses going cross-country.

Some riders use it with just a single pair of reins attached as above, but this isn't recommended. A second pair of reins should be attached to the bit rings in the same way as they are fitted to an ordinary snaffle, so the rider can employ either an ordinary snaffle action or the gag mechanism. If the gag action is used all the time, the horse tends to become impervious to it—and if by any misfortune one of the running cheekpieces breaks, the rider is left with no brakes and with steering only on one side.

The three-ring snaffle is sometimes referred to as a three-ring gag, but it doesn't have a true gag action. It works like a lever, applying poll pressure and tilting the mouthpiece forwards and up. The reins can be positioned on either of the rings and the lower they are set, the more leverage is imposed.

Although this is a bit that some riders like, because they feel it gives them more control, a lot of horses dislike it and will raise their heads and hollow their backs to try and get away from it: not a reaction you want at any time and certainly not when jumping. If you have to use it, try putting the reins on the top ring.

Double Bridles

Although a double bridle looks to be—and sometimes is—a real mouthful for a horse, the two bits can facilitate the ultimate communication between a skilled rider and a well-trained horse who both know what they're doing. In simple terms, using the bradoon rein asks the horse to raise his head and come back to the rider, while using the curb rein asks him to lower his head.

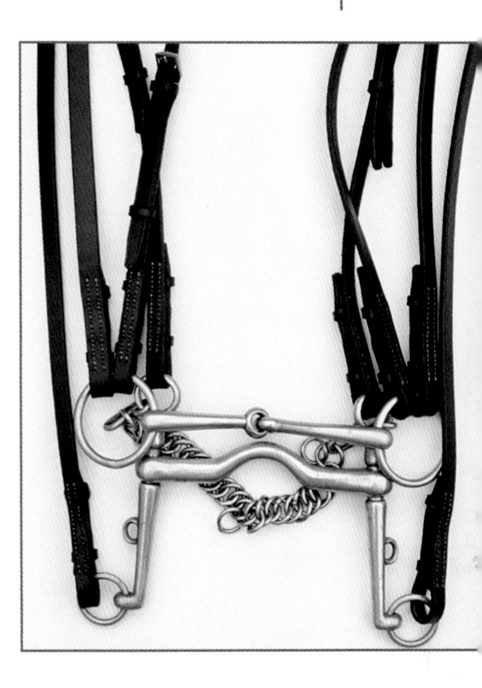

Problems arise if the horse or pony literally doesn't have enough room in his mouth for two bits or if the rider doesn't understand how to use a double bridle—for instance, if it is used in a mistaken attempt to give a more "advanced" head carriage than the horse is ready for, or to supply extra braking power. Manipulating the two pairs of reins attached to separate mouthpieces requires dexterity and quick reactions.

The bradoon, which is attached to a separate sliphead, should be a quarter of an inch wider than the horse normally needs to allow room for the curb and should be adjusted at the same height as an ordinary snaffle. Inside the horse's mouth, it rests on top of the curb and from the side, the curb should look as if it is sitting slightly below the bradoon. The curb attaches to the normal bridle sliphead.

The curb chain should come into contact with the curb groove at the back of the horse's jaw when the curb cheek comes back to an angle of 45 degrees. It can be made from metal links, or from leather or elastic to give a less definite feel for a particularly sensitive horse.

To be traditionally correct, you should also use a lipstrap, which attaches to the tiny loops on the curb cheeks and passes through the fly link. Its purpose is to ensure that the curb chain isn't dangling when one end of the chain is unhooked before removing the bridle. However, many riders no longer bother with one.

To adjust a curb chain, whether used with a double bridle, pelham, or kimblewick, hook it on to the offside curb hook, then twist it in a clockwise direction until it lies flat and the fly ring—the loose ring in the center of the chain—is at the bottom. Next, hook the flat chain on to the nearside hook at the appropriate length and if there are more than two links hanging down, hook them over the top for neatness.

Pelhams and Kimblewicks

Although pelhams and kimblewicks are less subtle than a double bridle, they are easier to fit and use and many horses and ponies go well in them. Pelhams can be used in show classes in place of a double bridle and both can be used for all disciplines except dressage (though there are some organizations who permit their use in their own tests).

Within this group, a pelham used with two reins gives the clearest and most effective communication. The top ring of the bit is designed to be the equivalent of the bradoon, though with some poll pressure, and the bottom ring, which applies more leverage, to equate to the curb. Although the action is far less precise than that of a double bridle, it suits many combinations. A rider

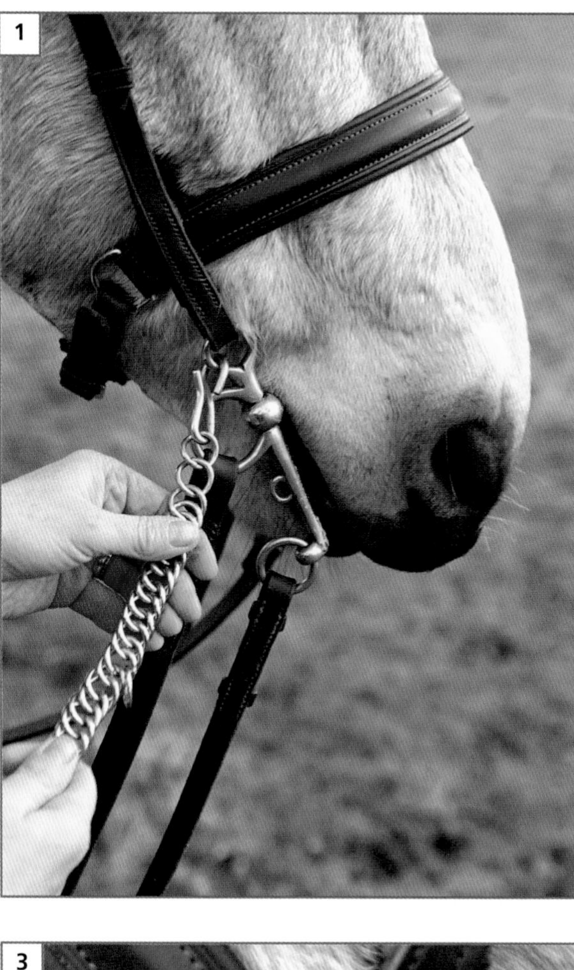

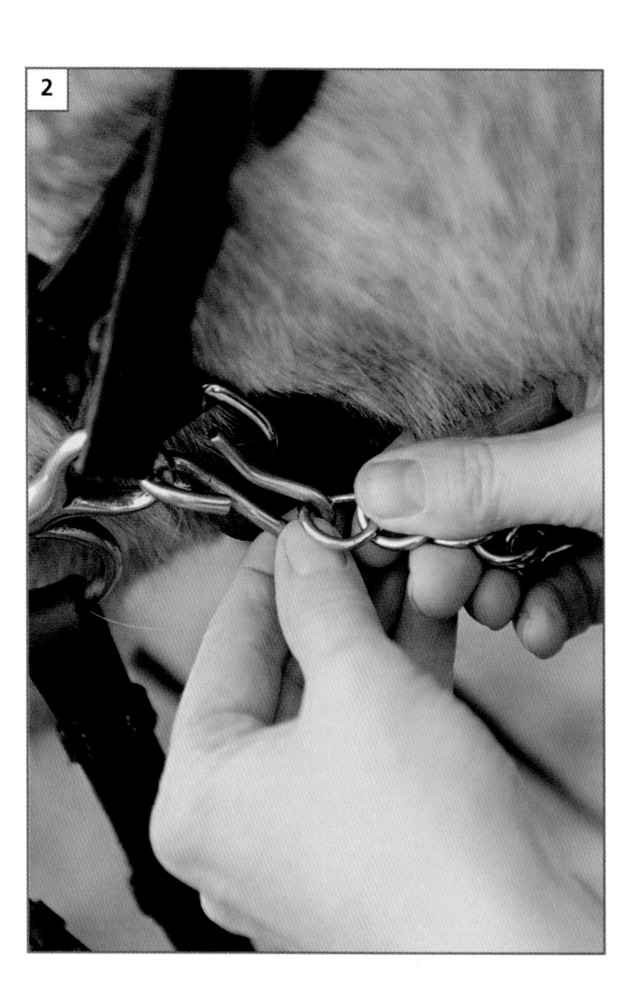

may gain more confidence when riding an onward bound horse or pony in a pelham and be able to sit in a more balanced position and use lighter rein aids. In turn, a horse will often be more comfortable, because he is not engaged in a pulling contest with his rider, and will carry himself more easily.

Pelhams, like snaffles, are available with numerous mouthpieces, but the most popular is the mullen mouthpiece. This can be made from metal, metal covered with hardened or soft rubber, or nylon. The curb chain can be fitted in the same way as on a double bridle, but is more likely to stay in the correct place if it passes through the top bit ring on each side.

Riders who find it too difficult to use two pair of reins can use a single pair with pelham roundings. These are leather couplings which join the top and bottom rings of the bit to form loops on to which the reins are fastened. They give an action somewhere between that of the top and bottom rein, but offer far less subtlety.

Kimblewicks are used with one pair of reins and again, are available in a variety of mouthpieces. The theory is that if the rider's hands are lowered, more curb action is employed, but in practice, a kimblewick is not capable of employing leverage unless it has two slots in the cheeks and the reins are fixed to the bottom ones. However, some animals go well in them.

Nose Pressure

Both bitless bridles, halters designed for riding, and combination bits such as those made by Myler act via nose pressure; the latter also obviously act on the control points of the mouth. Some trainers like to start a young horse's ridden education without a bit and as they will already be used to responding to nose pressure from being led in a halter or headcollar, there is logic in this system.

There are many designs of bitless bridle and the simplest are the scawbrig and sidepull. The scawbrig comprises a strap with rings at each end which goes across the front of the nose, with a back strap which goes through the rings and connects to the reins. This means that when rein aids are applied, it can act on the front, back, and sides of the horse's face.

A sidepull is, to all intents and purposes, a bridle with a cavesson noseband which has rings at the side to take the reins.

The English hackamore has short metal arms and a curb strap or chain. It has a more definite action but is not as potentially severe as the German hackamore, which because of its long arms can exert a lot of leverage. Dr. Cook's bridle

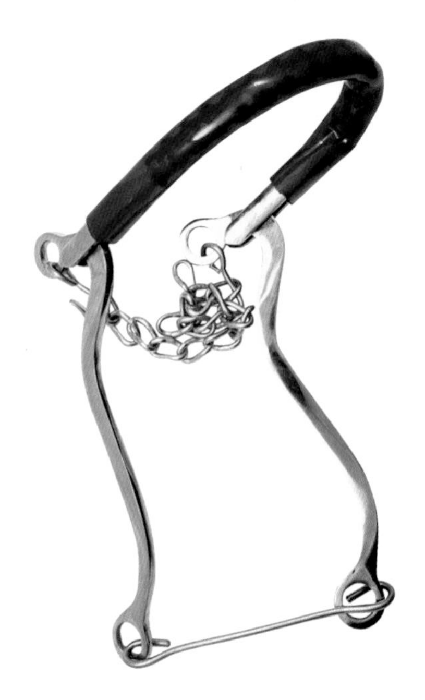

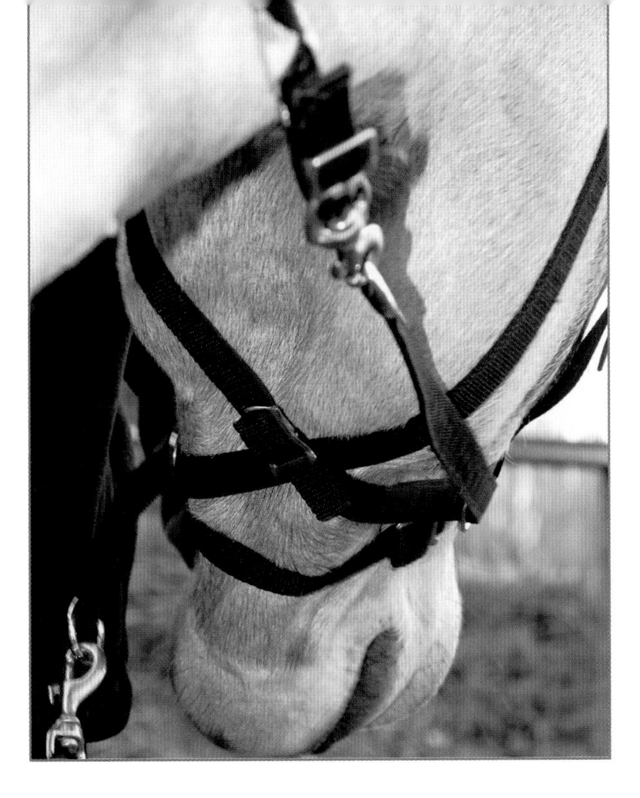

is designed by American veterinarian Dr. Robert Cook and is created to spread pressure over a wide area.

While it is perfectly possible to school a horse to go as well without a bit as with one, you should get advice on the design to choose and how to fit it from someone who understands its use. You should also ride in a safe, enclosed area until you are used to communicating with your horse this way.

Combination bits are often used successfully on strong horses, particularly by event riders going cross-country. The reason is not that they are severe, but rather that they have a kind dual action by working on the nose as well as the mouth. Neither bitless bridles nor combination bits are allowed in dressage tests.

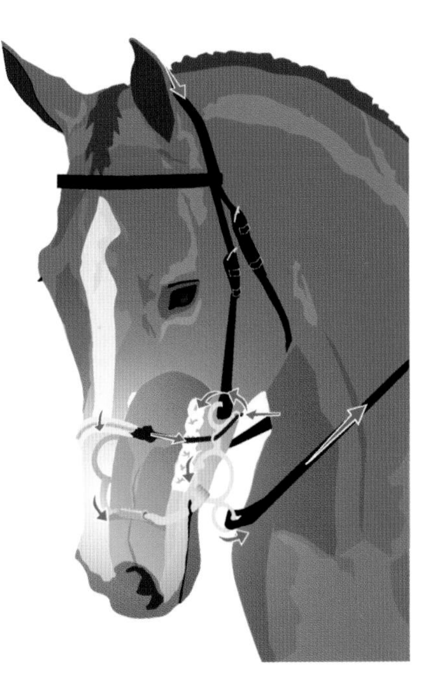

Martingales and Breastplates

Many riders consider martingales to be essential when jumping and some people prefer to use them all the time—though again, they are not permitted in dressage tests. Certainly they are a useful backup; apart from giving extra control when needed, the neckstrap can come in useful as an aid to slowing down (see the section on training aids) or as a grab handle in an emergency.

While you might prefer to use a martingale whenever you are hacking out or jumping, it's a good idea to try riding without one when you are schooling on the flat in a safe area. If you find that your horse immediately carries his head too high, or has an unsteady head carriage, then you've uncovered a weakness in your riding, his education, or both. If he goes equally well without one as he does with one, give yourself a pat on the back.

However, there is one situation where it can actually be preferable to fit a running martingale, which is the commonmost type in use and acts via the reins. If the horse is being ridden by a novice rider or someone with poor balance who tends to rely on his or her hands for balance, a running martingale will act as a buffer between the rider's hands and the horse's mouth by minimizing the effects of inadvertent little jerks on the reins.

The bib martingale, which has a triangular piece of leather joining the straps that link to the reins, has a similar action to the running one. Originally designed for horses who try and grab the martingale straps, it puts a little more weight on the reins because of its extra bulk.

The standing martingale is the only one which has

no direct action on the horse's mouth, as it fastens to a cavesson noseband or to the cavesson part of a Flash. Some trainers don't like it, because they feel a horse may set himself against it and build up muscle on the underside of his neck—just where you don't want it. However, others disagree and say it helps a rider communicate with a young horse who is starting his ridden education by acting on the control point he is used to.

Whatever type of martingale you use, there should be a rubber stop at the junction where the neckstrap and that running to the girth meet. This prevents the martingale looping between the horse's front legs and possibly causing an accident. With a running or bib martingale, stops should also be used on the reins to prevent the small rings sliding too far down and getting caught up on the rein buckles or small bit rings.

A martingale neckstrap should be adjusted so you can fit the width of your hand between it and the horse's neck. To check whether a running or bib martingale is at the correct length, hold the reins in the normal position when mounted. When the horse's head is in the correct position, the straps should not exert any downward pressure; there certainly should not be a bow in the reins.

A standing martingale should be adjusted first of all so that when fastened to the noseband, it can be pushed up to follow the line of the horse's throat. You may need to shorten it, but it should never be so short that it is tying down the horse's head.

A breastplate or breastcollar/breastgirth gives extra security to your saddle by preventing it from slipping back, especially when you are jumping. It should never be used to try and compensate for a badly fitting saddle, only as a backup to one that fits as well as possible. All aim to do the same job, but vary slightly in design.

Possible drawbacks to traditional designs are that if they are fitted just slightly too tight, they may restrict the movement of the horse's shoulders. New ones have more adjustment and attachment points.

Western Tack

Although Western-style riding has always had many devotees, over the past few years its popularity has soared and spread worldwide. Just as English-style riding has its own competitive disciplines, so does Western, and just as, say, a dressage rider will choose different tack from a show jumper, so a trail rider's tack will differ from that of a rider specializing in top level reining.

The construction of a Western saddle is basically similar to that of an English-style one, but different terms are used. For instance, a girth is called a cinch and many saddles are double-rigged with two cinches for extra stability. All Western saddles are designed for particular jobs, whether that be for a stock hand spending long hours in the saddle checking fencing or a wrangler cutting out cattle from a large herd.

There is as much skill in fitting a Western saddle as an English-style one. Trees are available in different profiles to suit different types of horse and as with an English-style saddle, it's important that there is clearance over the withers and along the back and that the saddle distributes the rider's weight evenly.

Many English-style riders worry that the curb bits used by proficient Western riders on schooled horses are severe. However, the styles of riding are very different; Western riders hold their reins in the left hand only and use only the weight of the rein, so a curb bit used in this way gives delicate signals. Most trainers educate young horses in a bitless bridle or a snaffle and only introduce a curb when the horse's education has reached a high level.

Training aids

In theory, any horse or pony can be schooled to go correctly using nothing other than a simple snaffle bridle. In practice, this isn't always possible, perhaps because the horse has been allowed to go incorrectly for so long it has become a habit—or it may have never been educated properly in the first place.

This is when a suitable training aid, used and fitted correctly, can help you show him what is wanted. There may also be occasions when using a training aid can bring about a breakthrough in communication or, when used under veterinary supervision, help in a rehabilitation program. There is no such thing as a magic gadget or a quick fix, but sometimes, using the right equipment means that the horse can be asked questions and rewarded for his response without any fear of imprecise timing.

Although some riders and trainers may talk disparagingly about gadgets and say they never use them, it's unlikely that this is the case. A training aid can be as simple as a neckstrap; in fact, it can be defined as anything that can be used as an aid to training. That definition is important; it is an aid to, not a substitute for, correct training. There are so many products on the market, all designed to influence a horse's posture, that it would take a whole book to list them and explain their actions. A few of the simplest to fit and use are shown here, but before rushing out to your local tack shop, be aware that there are golden rules for the use of any training aid.

If you are having problems with your horse because he is resisting your aids, first check that he is comfortable and does not have problems in, say, his mouth or back. This will usually involve getting him assessed by your vet and/or equine dental technician and getting the fit of your saddle checked by a knowledgeable fitter.

All equipment should be introduced carefully and fitted loosely to start with. This gives your horse time to accept the feel of something different; never take him for granted, even if he is an older, experienced animal. For instance, some horses are worried the first time they feel a lunge rein round their quarters.

Training aids should only be used with some form of simple snaffle, not with a lever or gag snaffle or with a bit that has a curb action.

If you have never used the equipment before, try and get help from someone who uses it successfully. Many manufacturers supply instructions for fitting and use and even on DVDs; study these carefully. Assuming you can work it out is like driving off in a new car without reading the handbook—by the time you realize you're in trouble, it's too late!

Anything that asks a horse to work in an unfamiliar way means he will be using unfamiliar muscles, so use a training aid for short periods to begin with. Often, five minutes on each rein will be plenty. Don't use it every day—adopting a new posture is hard work, as anyone who has lessons in yoga or Pilates will know, and muscles that have worked hard need to recover.

Some training aids are used for lungeing, some when riding, and some can be used for either. Most of the ones designed for use when riding should only be used on the flat, not when jumping.

A training aid is a means to an end, not an end in itself. If, after a reasonable time, your horse reverts to his old way of going as soon as the training aid is removed, you need a different approach. Finding a good trainer is always the best start!

Training aids only help if coupled with correct riding and/or lungeing. While a training aid might encourage your horse to carry himself better, you can't expect to get on him and for everything to happen by magic.

While many training aids can be used successfully on children's ponies, they should only be fitted and used under the supervision of an experienced adult.

Down to Basics

Sometimes, the best ideas are the simplest. In the case of training aids, this is certainly true.

A neckstrap, which can be either a stirrup leather buckled round the horse's neck or the neckstrap of a martingale or breastplate, offers a kind and effective way of teaching a horse to slow down, halt, or half halt. You should be able to slip the fingers of one hand underneath and give a pull and release on it at the same time as using your bodyweight and other aids.

Side reins are used when lungeing and encourage a horse to accept the bit and yield to its action. Get expert help in fitting them at the correct height and length for your horse's stage of schooling, but above all, don't fit them too tightly.

The elastic schooling rein or bungee rein encourages the horse to lower his head by applying gentle poll pressure when he raises his head too high and can be used when riding or lungeing. It is a long elastic rope with clips at each end which goes over the poll, through the snaffle rings, and then clips to the girth, either under the belly or at the sides. The former fitting gives a more definite action.

This rein—and in my view, every other training aid—is far preferable to draw and running reins. Draw reins fasten to the girth under the belly and running reins are attached to the girth at the sides. Both pass through the bit rings and go back to the rider's hands.

The problem with both is that they rely on the rider's hands to release their action the moment the horse obeys it. Quite honestly, few have such finesse and it's all too common to see horses being ridden with their heads forced down. When there are so many kind and effective alternatives, why use them?

Boots and bandages

Some riders fit protective boots on their horse every time they ride while others believe that they do not need to be used when hacking or trail riding. It's an individual decision and you need to factor in considerations such as the way your horse moves. In general, it's recommended that they are always used:

- When lungeing, because the horse is worked on a relatively small circle and may find it harder to keep his balance consistently.
- When a horse is young and/or unbalanced and more likely to trip or knock himself.
- When a horse's natural movement means he is likely to knock himself—for instance, if he brushes (knocks one leg against its partner on the opposite side).
- When jumping or doing fast work.
- When traveling, unless you prefer to use bandages over padding.

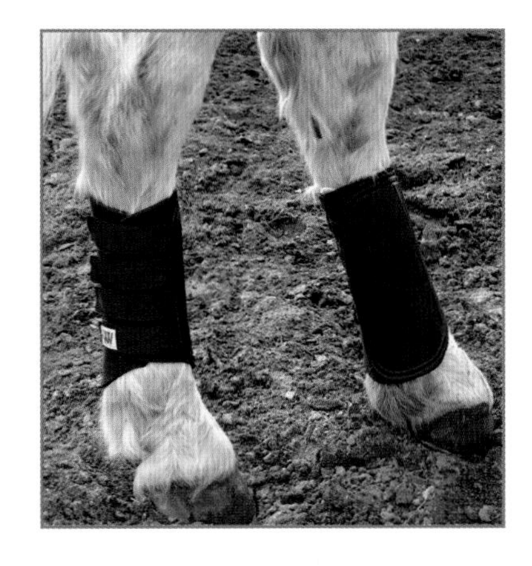

There are many types of boots, but the ones most commonly used when working a horse are brushing boots, tendon boots, fetlock boots, and overreach boots. Brushing boots protect the area from just below the knee or hock to the bottom of the fetlock joint and are a good all-around choice.

Tendon boots are used on the front legs and are reinforced in the part which covers the tendon, thus helping

to protect the horse from injury if he strikes the back of his front leg with the toe of a hindleg. Fetlock boots, as their name suggests, protect the fetlock area and overreach or bell boots are meant to minimize the risk of an overreach injury, when a horse strikes into the heel of a front foot with the toe of a hind one.

Many designs are made from lightweight but impact-resistant materials and are easy to wash. It's important to keep them clean, because if dirt is trapped underneath it may cause rubs and skin infections.

To maintain even tension when putting on brushing and tendon boots, fasten the center strap first, then the top one, followed by the bottom strap, and any others. When you take them off, undo the bottom strap first and work upwards, to minimize the risk of the boot flapping around the horse's leg and perhaps startling him.

Horses' legs should be protected when they are traveling in a trailer. The exceptions to this rule are broodmares and foals traveling together. It would be potentially dangerous to put boots or bandages on foals, as they travel loose.

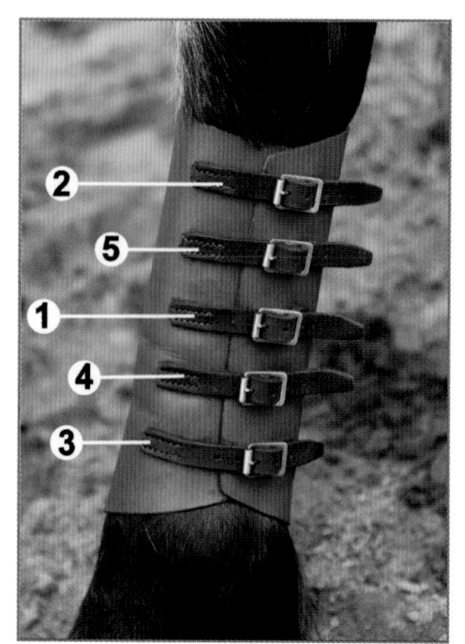

Well-designed travel boots are quick and easy to put on, but will only stay in place if they are the right size and suit the proportions of the horse's legs. Fasten and unfasten the straps in the same order as with brushing and tendon boots.

Leg Bandages

Leg bandages can be used to give protection while traveling and to add warmth or support in the stable when needed. They can also be used when exercising, but it takes great skill to apply them so they stay in place in this situation without causing unwanted and potentially harmful pressure. For this reason, most riders opt for boots.

The advantage of bandages for traveling is that if they are correctly applied, there may be less risk of slipping. The disadvantage is that they take longer to put on and remove and if your horse comes off the ramp bouncing around with excitement, removing them may be easier said than done.

One way around this problem is to tape over the bandage fastenings or stitch the end in place. You can leave them in place to protect the horse's legs while you work him in and remove them when he settles—if necessary, replacing them with protective boots.

The golden rules for using bandages are:

- They must always be applied over padding. This can range from Gamgee or felt-like materials to gel pads.
- Before you start, make sure the bandages are rolled up so that you finish with the fastenings on the outside of the fabric.
- Bandage from front to back to avoid putting pressure on tendons and ligaments and keep the pressure even as you bandage down, then back up the leg.
- Fastening tapes or ties should rest on the outside of the leg, at the side, so they do not put pressure on the tendons or cannon bone. They should not be tighter than the bandage, or they will cause pressure points.
- The bandage should be tight enough to stay in place without gaping, but you should be able to slip in a finger at the top.

Applying bandages takes a lot of practice, so try mastering the skill when you're not under the pressure of time limits! Although the traditional method is to wrap the padding around the horse's leg and hold it in place while you apply the bandage on top, it's easier to line up the bandage along the padding and apply the two together.

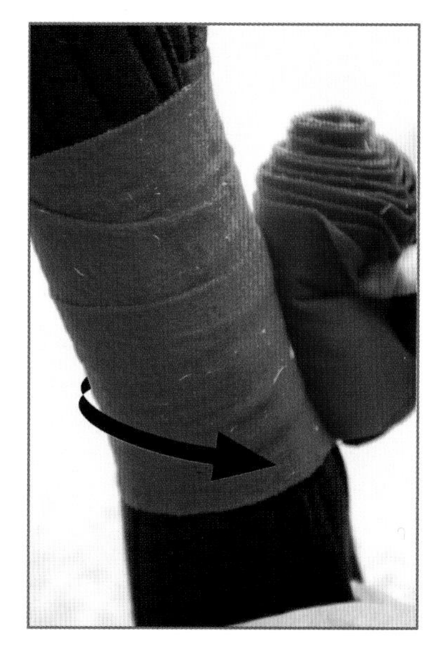

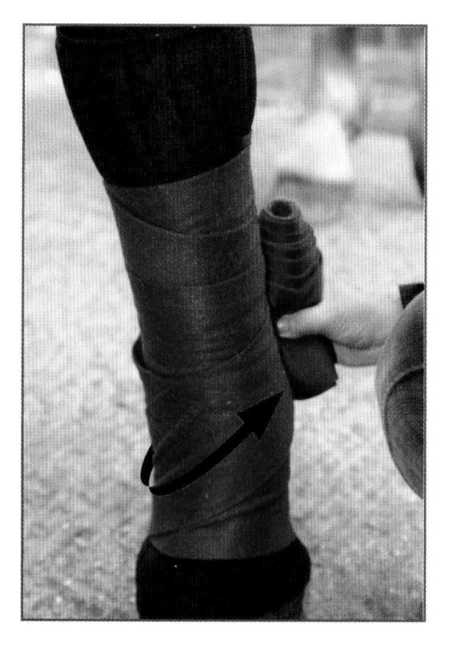

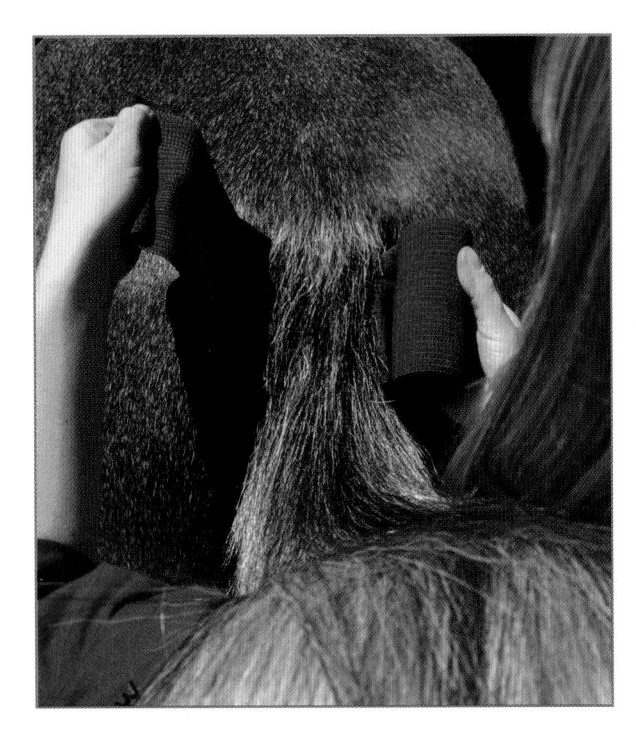

You will also need to practice putting on a tail bandage, which can be used when traveling, to minimize the risk of rubs, and for short, regular intervals to keep a pulled or shaped tail neat. Again, first make sure it is rolled so that the fastenings end up on the outside, then unroll a short section, and place it under the dock.

Leave a loose end protruding at the top, then make your first wrap, fold down the loose end, and bandage over it. This will help prevent the bandage from slipping. Next, bandage down with even, not too tight pressure until you reach the end of the dock, then wind the bandage back up again.

Fasten the tapes at the same tension as the bandage and tuck in the ends. If you fasten them in the bottom third of a wrap, it leaves enough material to fold over them and keep them out of the way. Finish by bending the tail gently into its natural position.

If your horse has a light colored tail and you want to finish a journey with it clean and sparkling, either apply a second bandage below the first or cut off the leg and foot from an old pair of tights to make a stretchy fabric tube. Pull it up over the tail and bandage over the top. You can buy "tail bags" that work on the same principle, but these are usually bulkier and some horses dislike them. They are also more expensive!

If your horse leans back against the vehicle and rubs his tail despite the application of a bandage, use a tailguard on top. This is a padded wrap that encases the top of the tail.

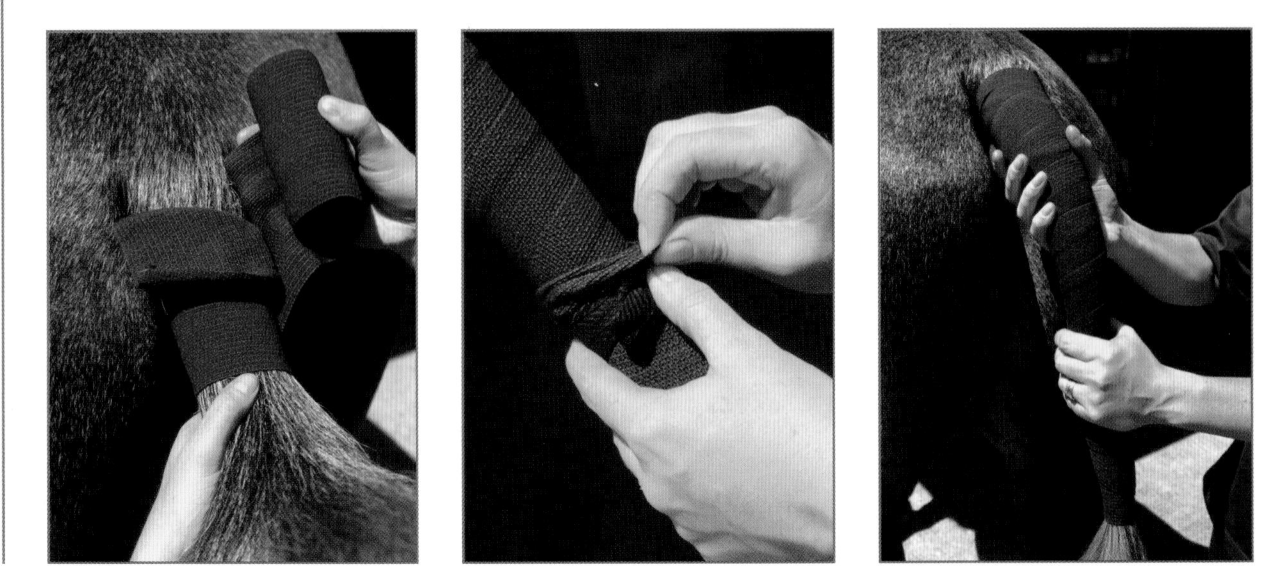

Blankets

There are so many types of blankets available that it's quite easy for a horse to have a wardrobe as large as his owner's. Fortunately, it's possible to keep spending at a sensible level by starting with essentials and adding extras later on if you feel they would be useful.

There may be cases where blankets are not needed at all and there is one school of thought that believes a horse's coat is the best insulator available. If he has good shelter, grows enough coat to stay warm and can be worked when his winter coat is in without getting hot and sweaty, this may well be true. However, being able to create such a setup is impossible for many owners.

While blankets may be used as much for the owner's convenient as for the horse's benefit—for instance, to keep him clean and dry enough to ride in bad weather—there are times when they are essential for his well-being. For instance, a horse or pony who suffers from sweet itch can be helped enormously by protective clothing that prevents the midges responsible from biting him, but does not cause him to overheat.

A horse who grows a thin winter coat may be too cold to withstand winter temperatures without discomfort and weight loss and one who is elderly or suffers from stiffness will certainly appreciate the protection a blanket can give. Another common scenario is the horse who must be clipped to allow him to work without the risk of him sweating and catching a chill. When he isn't working, he will need a blanket to compensate for the hair that has been removed.

If a blanketless lifestyle isn't suitable for your horse and you're on a tight budget, start by buying two lightweight,

breathable, washable turnout blankets that can be used outdoors and in the stable. You need two so that if one is damaged, you have a spare while it is repaired. If you live in an area where the winters are harsh, lightweight blankets won't be enough and you will need ones that provide more warmth.

When biting insects are a problem, fly sheets made from fine but reasonably tough mesh help keep a horse comfortable. If he suffers from sweet itch, you need a specialist design, as not all fly sheets are defense against the Culicoides midge.

When you can afford it, invest in a thermal blanket made from fabric that wicks moisture from the horse's coat through to the outer layer of fabric. These can be put on a

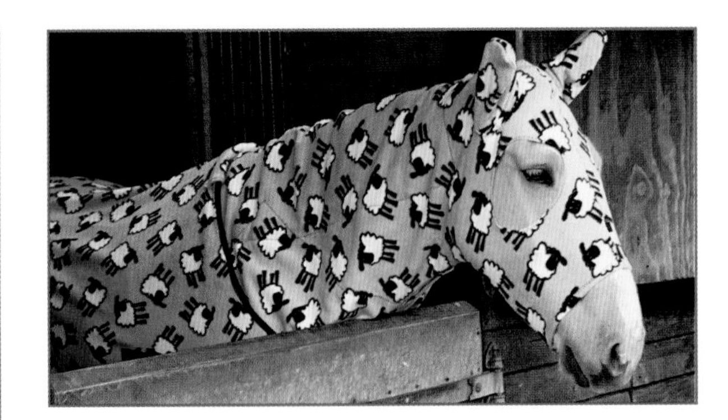

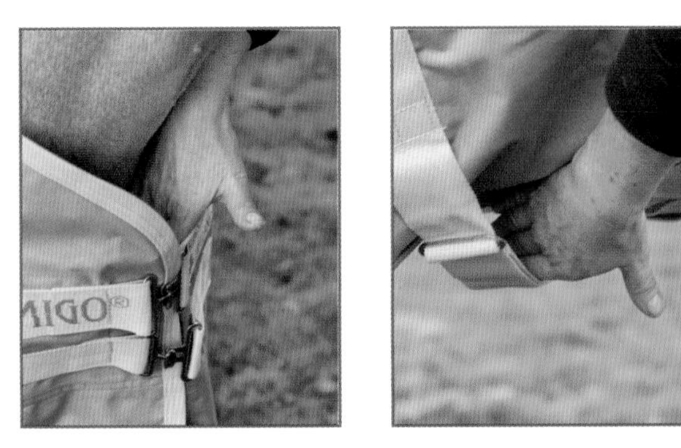

wet or sweating horse and mean that as he dries off, he stays warm. You can also get turnout blankets with the same sweat-wicking properties.

In summer, a fleece cooler will do the same job and you may also find a summer sheet useful. The latter are thin, cotton blankets which help to keep flies and dust off stabled horses. Both can double up as travel blankets in hot weather and can also be used underneath bulkier blankets. It's easier to wash a lightweight sheet than a heavier blanket.

In cold weather, particularly when it's also raining, an exercise sheet that fits behind or under the saddle will keep your horse comfortable when you're hacking out, or waiting to compete. Stretch hoods and bodies help keep mudlarks clean and some owners use them to inhibit the growth of the winter coat; this may allow you to get away without clipping your horse, or take off less hair than would otherwise be the case. Few horses object to hoods, but they must be fitted carefully and in safe surroundings. Ones with zips under the throats are easier and safer to put on.

Sizing and Fitting

A blanket will only do its job if it fits your horse. That might sound obvious, but a lot of people buy blankets that are too large in the belief that they will give their horses more protection. They won't, and they will be more likely to slip and cause rubs and pressure points. When a blanket is secured in place, it should start just in front of the withers and reach to the top of the tail. The tail flap, if present, will extend farther down the dock. To find out what size to buy, measure your horse from the center of his chest, along his body to the end of his quarters.

A blanket will only stay in place if the securing system is adjusted correctly. The commonmost system is a chest fastening coupled with cross surcingles, but blankets may also have legstraps or under-belly harnesses.

When the blanket is fastened, you should be able to fit a hand's width between it and the horse's chest to allow room for him to be comfortable while grazing. Cross surcingles should cross under the belly and again, there should be a hand's width between each one and the horse.

Front legstraps aren't recommended and to be honest, shouldn't be needed. Detachable rear legstraps may be fitted on some turnout blankets and unless the manufacturer states otherwise, are linked so that there is—you've guessed it—a hand's width between each one and the inside of the horse's hindlegs. Take the left-hand strap and pass it around the horse's hind leg, then clip it to the left-hand side of the blanket. Then pass the right-hand strap through the left hand one, take it around the horse's leg, and clip it to the right-hand side of the blanket.

If the horse isn't used to leg straps, fit them one at a time in safe surroundings and lead him around so he gets used to the feel of them. Otherwise, they may come as a bit of a shock and you could find yourself heading for his field rather faster than you intended.

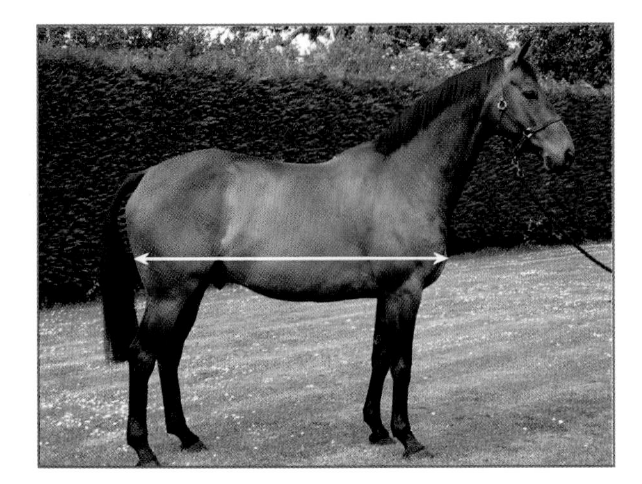

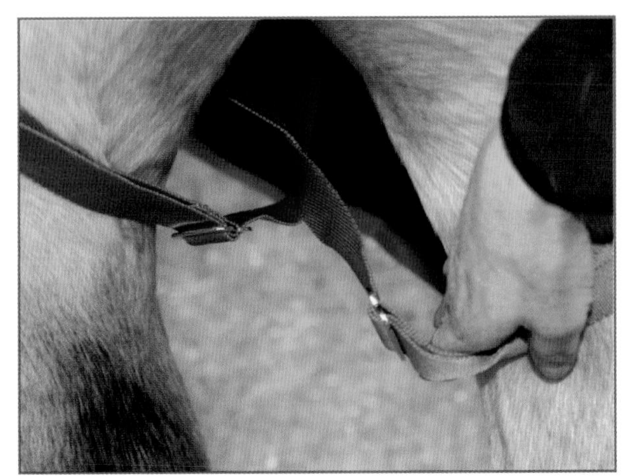

Staying in sight

If you ride on the roads—or in the woods during the game season—it's sensible to use fluorescent, reflective equipment so that you can be seen from all directions. Research has shown that the most effective combination is for the rider to wear a highly visible tabard or Sam Browne belt and the horse to be fitted with highly visible brushing boots or leg bands.

In daylight hours on the roads, pink or yellow equipment shows up well. During the hunting season, orange is the recognized color for hunters to look out for.

Riding on the road

Most horse owners will, at some point, have to ride on public roads. As even on the quietest routes you're likely to encounter all types of traffic, it's important to ride as safely as possible. Many drivers will be considerate, but you may be unfortunate and encounter some who don't understand that even the quietest horse can be unpredictable. To ensure your own and your horse's safety, and that of other road users, remember these guidelines:

- Always wear a riding hat to the current highest safety standards, correct riding boots, and high-visibility wear.
- Make sure your horse's tack is in good repair and correctly fitted and adjusted.
- Always ride on the appropriate side of the road to match the direction of the traffic—for example, on the right-hand side in the U.S.
- While there may be situations where it's necessary or appropriate to ride in single file, an experienced horse ridden on the outside of an inexperienced or nervous animal will provide an effective shield against the sight and sound of traffic.
- Do not ride on the roads at night, and try to avoid riding when visibility is poor.
- Do not ride on the roads when ice and snow are present. The surface may be unsafe.
- Always be considerate of and courteous to other road users. If a driver slows down and gives you plenty of room, acknowledge this with a smile and a nod of the head. This is safer than taking your hand of the reins to signal your thanks, as by doing so you're reducing your control over the horse.
- Always look behind you before moving out into the road to pass something or someone, and if appropriate make a hand signal.
- To signal that you wish to turn left or right, hold the reins in one hand and stretch out the other so that it's level with your shoulder.
- To ask a driver to stop, raise your hand with the palm outstretched towards the vehicle. To ask a driver to slow down, wave your arm up and down.
- Always tell someone the route you intend to take and roughly what time you think you'll return.
- Always carry a mobile phone in case of emergency – but remember that it's dangerous and potentially illegal to use it while riding.

Cleaning up

Although life is too short to take your tack apart and clean it every day, as laid down in horsekeeping manuals of fifty years ago, it's important to keep it in good condition and good repair. A realistic approach is to carry out essential tasks every time you ride and give it a thorough cleaning once a week—with the provision that if it gets wet or muddy, you deal with it before it's put away.

The daily essentials are to make sure that there are no dirt or mud on parts of your horse on which tack will rest, as this can cause rubs and skin infections. For the same reason, girths, saddle pads, and protective boots should also be kept clean.

When you are finished with your ride, rinse off the bit in clean water or use a bit wipe. This prevents saliva and traces of food drying on the bit that could rub the corners of the horse's mouth. Wipes impregnated with leather cleaner are useful for giving your tack a quick spruce-up before the next time you ride.

If you arrive home with wet tack, clean off any mud with warm water. Don't soak it—though in really bad cases you may have to dunk it in a bucket of water and lift it out again—because leather that absorbs an excess of water is prone to stretch. Let it dry naturally at room temperature, but don't place it in front of a direct heat source or it will dry out too quickly and become brittle.

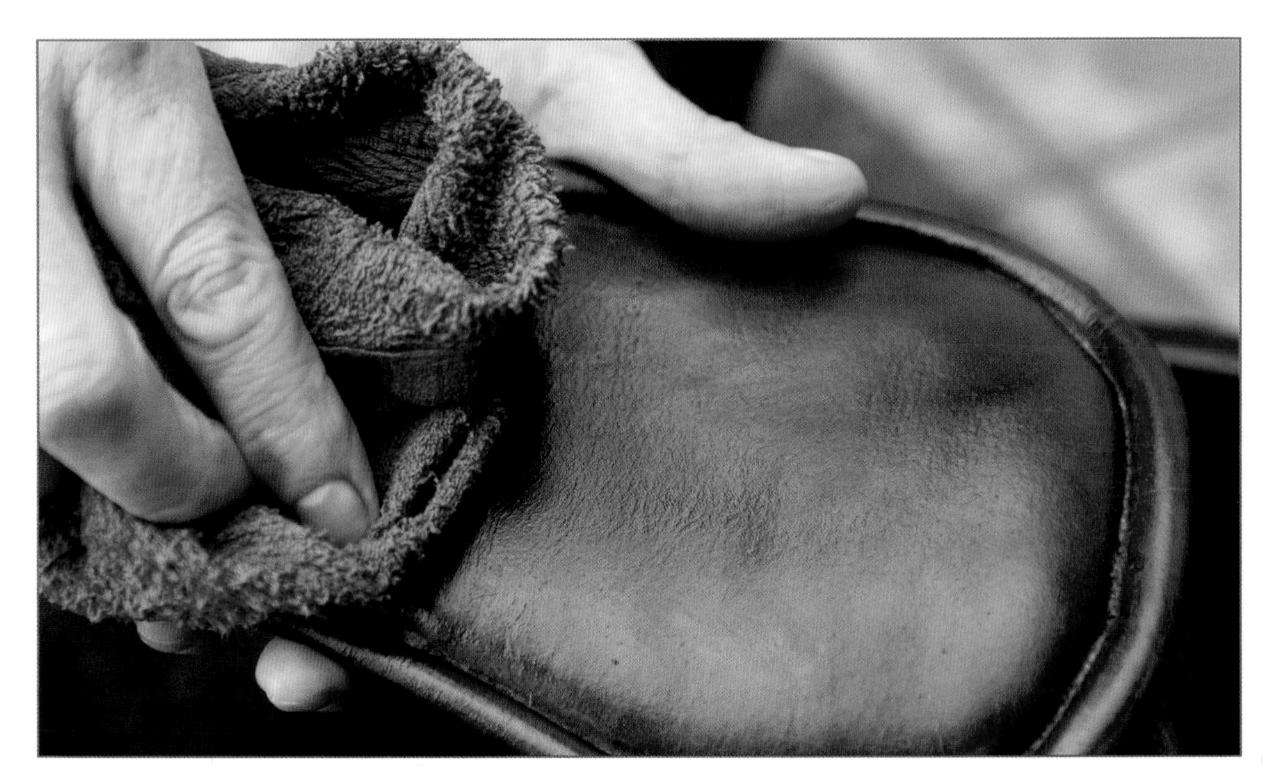

PART 6

BUYING A HORSE

If you've reached the stage where you're ready to buy a horse, prepare for a roller coaster ride. Finding the right partner is exciting but can also be nerve racking and frustrating. Above all, it's a responsibility that will not only place demands on your time and finances, but will also have an impact on other members of your family.

The previous sections in this book will have given you an insight into what's involved in finding a home for a horse, looking after him, keeping him healthy, and choosing the right tack and equipment. If you're planning to look after a horse yourself, either on a DIY or part livery basis, it's vital to appreciate the time and work involved and the best way to do this is to find a course at a local riding center that will give you hands-on experience under qualified supervision.

If you're a parent who is thinking of buying a pony for a child, you have to accept that the buck stops with you. Apart from practicalities like perhaps having to act as a transport service to the stable each day, you might have to take over checks and daily care when school commitments intervene—and if you have other children with different interests, can you ensure that no one gets left out?

Remember, too, that no matter how confident and competent a child is at handling and riding ponies, he or she should not be expected to do it alone. For safety's sake, an adult must be around in case anything goes wrong and to help with tasks that a child may physically not be able to manage. If you're going to become a "pony parent," you'll soon find yourself doing everything from buying feed to helping to tack up!

If you're an adult looking to buy a horse, be prepared to get the most out of every minute of the day and to work out—and perhaps rethink—some of your priorities. Most of us have to go without other things to pay for our horses and while you'll think nothing of buying him a new set of shoes every six weeks, you might not be able to be so generous with yourself! Also, will your partner/family/friends accept that some of the time you spent with them will now be spent with a horse? If you're lucky, they'll learn to share at least some of your interest—but it's no coincidence that a best-selling T-shirt bears the slogan: "He said it was him or the horse—we'll really miss him."

If you have any doubts, it may be best to start off by sharing a horse or pony with an owner who needs help or a contribution towards costs. This can work well for both sides, but it's always best to get a written agreement on who is responsible for what. You also need to be clear about how much and what type of riding you can do—for instance, if you are looking to share a horse, do you want the chance to compete and should you have lessons? If the horse's owner regularly has lessons with a particular instructor, it's an excellent idea to try and follow suit. This means you're following the same principles and someone who already knows the horse can help you build a relationship with him.

When you're ready to enter into a greater commitment, you need to decide whether you are going to buy, lease, or loan. The only real difference between leasing and loaning is that leasing involves paying a fee and is in many ways a rental agreement, while loaning means you will be responsible for the animal's keep, running costs, and any other agreed expenses.

Buying may seem to have the obvious disadvantage that you have to pay out a lump sum, but many people prefer to do this and be able to make all decisions about where the horse is kept, what activities he can be used for and so on. Also, the purchase price of a first or even a second horse or pony may be relatively small compared to the costs of keeping and equipping him. If you are tempted to try and take a horse on loan because you can't afford to buy one, you may well find that you can't afford to keep him!

However, there are many owners who offer horses on loan for genuine reasons—for instance, perhaps a much loved pony has been outgrown by one child in the family but will be wanted by a sibling who isn't yet ready for him. If you can reach an agreement on where and how an animal is to be kept and can accept that his owners will probably want visiting rights, it's a system that can work well for compatible loaners and borrowers.

You may also find that there are people who specialize in leasing horses and ponies, especially ponies. This means that you pay a lease fee in return for taking responsibility for the animal for an agreed period, usually a year. The fee often includes the use of the pony's tack and perhaps insurance (see later in this section) and hopefully means that you or your child will be matched with a suitable horse or pony.

Before entering into such an arrangement, get feedback from previous clients and try the animal as carefully as if you were intending to buy him, as explained later in this section.

Preparing to buy

If you decide that the time is right to buy a horse, you'll soon find that there are thousands for sale at any time. You'll find ads in local and national publications, on internet sites, and even on notice boards at local equestrian businesses. Don't forget word of mouth, either, as equestrian jungle drums can be very effective! In particular, children's ponies who are known to be safe and reliable often find new homes without being advertised.

When you're not looking seriously for a horse, it's fun to browse and imagine yourself riding off into the sunset on the 16.2hh five-year-old bay Thoroughbred gelding that according to the vendor's advertisement is perfect in every way and certain to make a top class competition horse. In reality, even if the horse matches up to the description (and you can't assume that it will) you would need to be an experienced rider with serious ability, serious ambition, and probably a serious amount of money to spend in order to consider such a purchase.

If you're looking for a first or even a second horse, buying the example above would be the equivalent of someone who has just passed a driving test buying a Ferrari—a disaster waiting to happen. It's important to buy a horse that will suit your current level of ability and fit in with your lifestyle, not one to suit the rider you hope you will become, and accept that as your ability and perhaps ambitions grow, you may have to find him a new home and move up to a horse who will take you to the next stage. This is even more important when it comes to buying a child's pony.

While it's difficult to lay down hard and fast rules about what sort of horse would suit a particular rider, there are useful guidelines to think about. If you or your child have been having lessons with an experienced instructor, ask for some advice about the sort of animal that would fit the bill. Think, too, about horses and ponies that you or your child have "clicked" with and see if there are common denominators in their height, personalities, and the way they respond to the rider.

A lot of people make the mistake of buying horses that are too big for them; sometimes, this is for no other reason than they think it makes them look good. But while a tall, fairly heavy rider wouldn't be happy with a small, light-framed horse (and vice versa!) most adults will be suited by a horse of 15hh—16.1hh.

Height isn't everything: a deep-bodied, 15hh cob can accommodate a longer-legged rider than a 16hh horse with a shallow girth. Another thing to remember is that adults can also ride ponies; the larger native breeds will carry riders up to about 5 feet 10 inches, and are versatile, athletic, and usually cheaper to keep than bigger horses.

The Fjord pony pictured here, who competes successfully in dressage with her owner, is capable of carrying a much heavier rider.

Age isn't as important a factor as some people think. It's often said that the magic age range for a first or second horse is between five and nine years old, when the horse should be old enough to have received a good basic education and gained some experience, but young enough to have many more years ahead of him. However, while you may be lucky enough to find a horse within that age bracket, don't discount those outside it.

A competent adult rider who can work under the guidance of a good trainer may find that a four-year-old with an exceptional temperament fits the bill. The advantage of this sort of horse is that he will be unspoiled; however, while he may be willing and trainable, he will also be inexperienced and you will need help to make sure he gains this in the right way.

Similarly, a horse in his teens who is sound and suitable for a novice rider can also be a good buy. However, it has to be said that there are many older horses who are not suitable for novices, either, because they are too sensitive in their responses or because they have learned too many evasions. Also, an older horse who is a competition schoolmaster will command a large price.

Ponies tend to have longer working lives than horses and many are still going strong in their late teens and twenties. Older animals, like older people, eventually start to feel their age and may develop conditions such as arthritis that need careful management—but with the right care and a suitable working regime, an old friend can be with you for many years.

Finding the right equine partner can be as individual as finding the right human one—it may take ages and you may have a lot of disappointments along the way, or you may strike lucky and find him or her the first time you look over a stable door! To give you some idea of what to keep in mind, here are a few scenarios.

The First Pony or Horse

While the first pony might not be the most expensive equine a parent ever buys, in many ways he's the most important. By first pony, we're talking about one to be ridden off the lead rein, so that the child rather than the handler is in control. The right one will give a child a lot of fun and build confidence, while the wrong one can put off a young rider for good.

First of all, make sure the pony is the right size and shape for its rider. Don't be tempted to buy one that is too big in the hope that your child will grow into it, because although you might be lucky, chances are that her legs will be too short to give clear signals and she won't have the strength or balance necessary to stay in control. For simplicity's sake, I'm referring to "her," but it could equally be "him."

The pony must also have a placid, equable temperament. Even though an adult should always be around to supervise, it must be happy to be handled, to have its feet picked out, and so on. It's no good having a pony who is a paragon to ride but threatens to kick every time your child wants to pick up its back feet.

A first pony doesn't have to be beautiful, but there are aspects of conformation that makes it easier for the child to ride. A small rider doesn't want a pony who is too wide, but when she's in the saddle there should be enough neck and body in front of her to give her confidence. If a pony is very narrow, or has such a short neck that its ears are virtually in the rider's mouth, she won't feel secure.

Another conformation fault to avoid is the pony whose back end is higher than his front, so he is built downhill. This means he is naturally on his fore hand and will find it easy to lean on the bit and pull the reins out of a child's hand—both of which are very tiring for the rider.

When he's ridden, the first pony should be obedient, but not super-sensitive. When the child gives a little kick, the pony should know what it means and obey. This may sound an anathema to adult riders who know to use pressure from the calves as an aid to go forward rather than a kick, but remember that small children don't have the musculature or strength to do that. However, the rider shouldn't need to keep kicking.

The ideal first pony will be obedient to ride and kind to handle, as well as being good to catch, load, shoe, and travel. He should be good in traffic and although there is no such thing as a bombproof pony, he should be as near to it as you can find.

Inevitably, the perfect pony will not be a youngster and will probably have filled the same role for several families

before you find him. As long as he is sound, you don't need to worry too much about his age. Equally inevitable, your child will eventually outgrow him and unless a younger brother or sister is waiting to take over, he will need to find another home. Even this becomes part of the growing up process: when you love your pony, it's a hard but necessary lesson to appreciate that the time has come to let him move on and give confidence to another rider.

The first horse should fit many of the specifications of the first pony—good to handle, sensible and obedient to ride, and willing without being silly or fizzy. You will probably find that an established all-arounder with a forgiving nature is a good choice.

A lot of riders fall in love with a horse or pony they have ridden regularly at a riding school and want to buy him. Even if this is possible, it might not be a good idea. Riding school animals get used to a routine and to being ridden in groups and while some adapt to a different lifestyle, others don't.

The All-arounder

The all-arounder is the horse that 99 percent of riders want. He should be a real Jack-of-all-Trades and capable of having a go at everything at the equivalent of Riding Club level—dressage, show jumping, cross-country, and, if he fits into a particular category and has no major blemishes, perhaps some showing as well.

The perfect all-arounder is a horse who is easy to live with; he has a nice temperament and can fit in with your lifestyle. He won't be a world beater (but then, how many of us could ride a horse that is?) but he will give you an enormous amount of fun. He should be good on the roads

and trails, both alone and in company, and shouldn't have any real hang-ups.

All-arounders can come in all shapes and sizes, but if you had to pick an overall type it would be the 15.2hh—16hh Quarter Horse or Quarter Horse cross, the other half being a cob or draft horse. The Thoroughbred cross Connemara and Thoroughbred cross Welsh combinations are often athletic but sensible. However, when you're looking for an all-arounder, keep an open mind. Again, cobs often fit the bill perfectly, as they usually have rhythmic paces and a natural jump, and depending on your height and weight, one of the large native breeds could also be ideal.

The Schoolmaster

There is a world of difference between the first horse or pony and the schoolmaster or mistress. The first is an obedient but not too sensitive ride, while the second is a horse who has been schooled to a high standard, perhaps in dressage or show jumping, and is straightforward enough to be ridden by someone starting out in a particular discipline and wanting to specialize in it.

The sort of rider who can benefit from a schoolmaster is the person who is competent and reasonably experienced and has already competed at basic level—perhaps someone who has done well with an all-arounder and decided that his or her ambitions lie in pure dressage. A true schoolmaster will help you learn how to fine tune your riding, because if you don't press the right buttons at the right time, you won't get the right results.

Many schoolmasters are horses who have reached a certain—and often high—level of competition but don't have the ability to go further. They are very hard to find and don't come cheap.

The Ride-and-Drive Horse or Pony

Driving has become enormously popular and the ride and drive animal, who is just at home in harness as he is under saddle, can be a real family friend. Depending on your family circumstances, you may want to look for either a smaller pony who can be ridden by a child and driven by a parent, or a large pony/small horse who can be ridden and driven by adults. Obvious candidates are partbred native ponies and cobs.

Unless you're an experienced driver, look for a horse or pony who knows his job. It's possible to have one who has been used not only for riding, but also broken to harness—and many people do this to give outgrown ponies a new job rather than selling them—but if you're starting off with an animal you don't know, you want one who knows more about driving than you do!

He must be good on the roads and fairly unflappable as well as being a reasonably well-schooled ride. You might have to compromise and accept that a good driving animal could have a slightly higher knee action than is ideal for riding, and also that his trot will probably be faster than is ideal for riding. If he has done more driving than ridden work, you'll probably also find that canter is not his best natural pace.

Despite that, some horses and ponies adapt to whatever job they are asked to do; when the harness comes out they go into driving mode and when a saddle is put on, they switch to riding mentality. Schooling can also make a big difference: some ride and drive horses will have concentrated mainly on driving and while safe to ride out, will not have much finesse. A good instructor will be able to help you improve this.

Ex-Racehorses

Why include ex-racehorses as a scenario? Quite simply, because there are a lot of them about and they are often offered relatively cheaply—which may make them an attractive proposition to someone on a tight budget.

Please don't consider a racehorse straight off the track if you are a novice rider. With correct management and schooling, they can become wonderful riding horses and many go on to be successful in competition—but when they first come out of racing, they definitely aren't for first-time riders or owners.

Of the thousands of horses bred for the racing industry each year, only a handful actually have successful racing careers—and in many cases, those careers are very short. Many don't actually race at all, either because they are too slow or because they develop soundness problems. The first doesn't affect a horse's ability to go on to another job, but the second is a big consideration.

Leg problems, particularly tendon strain, are common. Depending on the severity, tendon strain may not affect a horse's ability to compete in other spheres, but there will always be a weakness there.

While horses out of racing are individuals, they are bred for their speed and quick reactions. They are also broken and ridden in a different way from other horses; horses who are intended to race on the flat are broke as yearlings and trained and raced before they are physically mature. Even though their riders are lightweight, it is a lot for them to accept, both physically and mentally.

Racehorses are ridden in their daily work and in races by riders with very short stirrups, so do not understand what conventional leg aids mean. In many stables, they are also kept stalled for 22 or 23 hours a day and fed high energy feed with little forage, so they may also be prone to conditions such as stomach ulcers and to stereotypical behavior.

The good news is that they often have a wide experience of life and will be used to traveling and to traffic met on the way to the track. The lucky ones go to specialist stables approved by the racing authorities who reeducate them before finding them suitable new homes. This involves

Many ex-racehorses go on to be successful in other careers.

turning them out for a period to relax, then treating them as if they were unbroken youngsters. Anyone interested in buying an ex-racehorse is strongly recommended to find one via a specialist retrainer rather than directly from a racing stable.

The Rehabilitated Horse or Pony

There are many equine charities who take in horses and ponies whose owners can for various reasons no longer keep them—and also those who, sadly, are victims of cruelty and/or neglect. In some cases, these animals can be rehabilitated, if necessary, and put out on loan to approved homes.

Charities are very careful where horses go and expect potential borrowers to undergo riding and practical assessments. They also inspect the premises where a loan horse will be kept and make unannounced checks to make sure it is being kept and worked correctly.

This may sound a bit severe, but everything is done in a pleasant way. If you are considered suitable to loan a horse, you will have plenty of advice on hand from the staff who have looked after and perhaps rehabilitated him and if you form a good partnership, will experience enormous satisfaction.

The thing you have to remember is that many horses end up in this situation because they have a problem, either physical or behavioral. Thanks to the skill of the staff who look after and perhaps retrain them, they may go on to lead useful and successful lives, but their past experiences may limit the work they can do or the type of home they can be placed in.

If you are interested in the idea of taking on such a horse, find out more about the work of the different charities and study case histories of some of the horses and ponies they have put out on loan.

How to buy

There are four ways of buying a horse—from a private seller, from a dealer, through an agent, or at an auction. All have advantages and disadvantages from the buyer's point of view, but whatever avenue you explore, remember that buying a horse is not like buying a car. You're dealing with a living creature who has a mind of his own and though hopefully, you'll find the right partner, he will still need time to get used to a new home and a new owner.

Private Sales

Many people assume that buying from a private seller—which means someone who is not selling the horse as a business transaction—will mean paying a smaller price and that there is less chance of being taken for a ride. This isn't necessarily the case; while many private sellers are knowledgeable, sensible, and truthful, there are others who unfortunately are not. If they are selling a horse because they don't get on with him, they are unlikely to say so and you may also find that lack of knowledge prevents them from assessing and pricing their horse realistically.

The advantage of buying from a private seller is that if you're lucky, you'll be dealing with someone who wants to sell a horse or pony for a good reason—perhaps because he's been outgrown, or because the rider wants to move up to a higher level of competition—and wants to ensure that he gets a good home. If the seller knows the horse well, you should be given an honest description.

Trade Sales

Buying from a professional seller (colloquially known as a dealer or trader) means you'll be paying a good market price, but the advantages can be worth it. Apart from the fact that you may be able to see several potential purchases on one visit, dealers know how to assess riders as well as horses. A good dealer will tell you if a horse you want to buy isn't suitable for you and will steer you towards the type that would—not because he or she is necessarily kind-hearted, but because dealers rely on their reputations to earn their livings. If you take home an unsuitable horse, the word will soon get round who sold it to you—but if your new partnership is a success, you'll be a good advertisement for the dealer.

While many dealers sell a wide range of horses, others specialize in competition horses. These are the people to visit later on in your riding career, not when you're looking for your first horse.

Sometimes, private owners will offer horses and ponies for sale through professional sellers and trainers. There can be genuine reasons for doing so—for instance, if they don't have the facilities to show off a horse, or simply don't want to have to deal with potential buyers, sending him to a professional seller to be assessed and sold may be a sensible option.

Auctions

Only the very experienced should buy from an auction. There tend to be three types of auction sale—the top of the market performance sales, those organized by breed societies and organizations, and local sales.

While some offer good facilities for trying horses, others don't. Even if you can try a potential purchase, you have very little time to make up your mind and must be sure you understand the terms and conditions of sale. For instance, you might find that if it isn't stated in the catalog description that a horse is good to catch, box, and shoe, you can't complain if you get him home and find you can't do any of these.

Also, not all auctions offer the chance to have your purchase inspected by a veterinarian to make sure he is sound, though some have vets on site or specify that animals must be offered with a veterinary certificate.

Asking the Right Questions

Once you've got an idea of the sort of horse or pony you're looking for, you can start reading the ads in local and equestrian publications and on internet sites and the notice boards in local feed and tack stores. Some may sound too good to be true, and probably are: remember that sellers will make their horses sound as attractive as possible.

When you find animals you think might suit your purpose, ask a lot of questions before you go and see them or you'll waste a lot of time and money making useless journeys. It's always a good sign if the seller asks you lots of questions, too, because it usually means there is a genuine desire to find the horse the right home. Don't be a time waster; some sellers will be prepared to negotiate on the price, but don't arrange to see horses advertised for twice the size of your budget.

Questions to ask include:

- Is the seller a private owner or a professional seller?
- How long has the horse been in its current ownership?
- Is there documentary proof of the horse's age? Dentition is a good guideline but not an exact science, so documentary proof through breeding papers is useful.
- Has the horse been measured, or is the seller guessing at its height? Some guesses are more accurate than others!
- Will the horse do the job you want him for? Tell the seller exactly what this is. Never exaggerate your ability or experience; you might dream of competing at top level dressage or eventing, but if you know that riding club activities are a more realistic option, be honest.

- Does the horse show any signs of stereotypical behavior?
- Does he need any special management, perhaps because of allergies?
- Is he safe to ride on roads or trails both on his own and with other horses?
- What sort of bit and bridle is he ridden in? If a horse who is ridden English-style always goes in a simple snaffle, that's an indicator that he's fairly amenable. If he's ridden in a gag snaffle, Grakle noseband, and running martingale, you can assume he's not so easy!
- Is he good to load, travel, catch, and shoe?
- If you bought him, would the owner be prepared for him to undergo a veterinary inspection?
- Why is the horse for sale? Most sellers will volunteer this information and it's up to you to decide whether or not you think they are telling the truth.

Face to Face

The answers to the above and any other questions you ask should give you a good idea of whether it's worth looking at a horse. Remember that there's no such thing as a perfect horse or pony—but if you have to compromise on something, make sure it's a compromise you can live with.

Unless you're an experienced owner, it's a good idea to take someone whose advice you trust when you go to see a horse. However, as a professional such as your riding instructor will probably be expect to be paid for this service, you might want to make a first visit alone and if you like the horse, arrange to return as soon as possible. You might want to videotape the horse to show his conformation and movement and to get someone to video you riding him, but it's polite to ask the owner's permission before turning up with a film crew!

Assessing a horse for sale should be a logical process. Although it's difficult, you have to let your head rule your heart. Start by looking at him in the stable or the field; ideally, he should be in a stable and reasonably clean. If he's been dragged in out of a field and his legs are covered in wet mud, either the owner is very näive or is trying to hide problems. In either case, ask the seller to hose the legs off so you can feel the horse's legs and pick up its feet.

The first section of this book tells you about conformation and movement, and gives you a blueprint on which to base your assessment. At this stage, as long as there are no glaring errors, keep an open mind. If you find any lumps, bumps, or other blemishes, make a mental note to ask your advisor or vet's opinion if you decide that this is the horse for you.

While you have to be practical, you also need to ask yourself if you like this horse. Does he appeal to you and do you feel drawn to your first impressions of his character? This is a difficult one, because some horses are naturally people friendly, while others are more aloof and still more only open up when they are in a one-to-one relationship with an owner.

Be aware, too, of the way the seller handles the horse. A professional should handle him confidently and quietly, but if a private seller appears nervous or overly dominant, warning bells should sound.

If for any reason you have already decided that you don't want to buy this particular horse, say so, but in a nice way. It's much better to say that he's very nice, but not quite what you're looking for, than to tell the owner that you don't want him because he's got cow hocks and dishes.

So far, so good? Then ask the seller to tack him up and ride him for you. It isn't a good idea to get on a horse unless you've seen someone else ride him first; not only do you take the risk that the horse will buck or take off, you need to see what sort of ride he gives to someone who knows him.

Watching the horse while he's being handled and tacked up may give you clues to his temperament and attitude. Is he calm and friendly and does he move to one side if asked to? Or does he barge around, or bite or kick when his girth is fastened? If he associates the saddle with discomfort, it

could be that he has been handled inconsiderately, or it may be that it doesn't fit properly and is causing pressure on its points.

The seller should ride the horse for you at walk, trot, and canter, and, if appropriate, jump him. In many cases, this will be done in an arena, but not everyone is lucky enough to have such facilities and he may be ridden in a field. While an arena gives a safe environment, it's also nice to see a horse go well in an open space. Make sure you see him in all paces on both reins; the rider will need to warm him up, so just take in the horse's way of going. Are his paces calm and rhythmical and does he seem obedient?

If you intend to jump, ask the rider to show you how the horse performs. You don't need to see him tackle big fences even if the seller claims he is capable of this; for a novice rider, the height doesn't matter as much as his jumping style. He should be calm but willing and stay on a rhythmic stride approaching and going away from his fences.

By now you'll either be eager to try the horse or you'll have realized he isn't the one for you. Again, don't feel bad if you've come to the latter conclusion. Most sellers, particularly professional ones, would rather you said as much than wasted their time and put the horse through unnecessary work. If you think the horse would be too much for you, perhaps because he is very forward going, it's definitely better to say so rather than risk frightening yourself and perhaps upsetting the horse.

If you like what you've seen, ask to ride him yourself. Take your time and try not to feel under pressure, even if you feel a bit nervous about riding in front of other people. Make sure the girth is tight and that your stirrup leathers are at a comfortable length and start off in walk to get the feel of the horse. Try some turns and circles on both reins to establish your steering and ask him to halt.

For readers who haven't yet started learning to ride, riding on the right rein means you are going in a clockwise direction, either on a circle or around the school. Riding on the left rein means you are going in an counter-clockwise direction in a circle.

Once you feel you're speaking the same language, move forward to trot and again, ride turns and circles on both reins. Ask for transitions between paces, moving up and down between trot and walk. What happens when you ride him down the center of the ring, without the support of the fence—can you keep him reasonably straight? Does he feel equally easy to steer on both reins, or does he seem less supple on one than the other?

Unless a horse is well schooled and his rider is well balanced, riding on one rein often feels easier and smoother than riding on the other. It's something that can be improved, but if you're a novice rider, you don't want to feel that you can only communicate with the horse when you go in one direction.

Still happy? Then ask for canter, again on both reins. Start by trying a 65 foot circle on each rein, then go large (canter round the edge of the arena or in a straight line in the field.) Does he go up and down between paces happily and does he strike off on the correct leg each time?

When a horse canters, he takes a longer stride with one fore leg than the other and the leg which takes the longer stride is called the leading leg. For him to stay balanced, the right leg should be the leading leg when he is on the right rein and on the left rein, it should be the left leg.

If you want to jump and have reached that stage in your riding, it's a good idea to try the horse over a couple of small ones. Don't feel under pressure to tackle higher fences than you feel comfortable with, even if the seller showed that the horse was capable of it.

By now, you'll have a good idea if this horse could be the one for you. However, if you're going to hack out or trail ride, you need to make sure that he is safe in these circumstances. Many horses are more forward going out of the confines of an arena, but they should still be obedient—and while no horse can ever be said to be bombproof, one for a novice rider should be reliable enough to pass the sort of hazards you are going to encounter, particularly traffic.

Assessing this can be a little difficult, particularly if the horse lives in a different type of environment from that which you would take him to. For instance, country horses may be perfectly happy to ride past fields of sheep or cows but be unused to very busy roads, while town horses may cope with a relatively high volume of traffic going past but wonder what the funny looking woolly things are that you're asking him to go past. In general, a sensible horse will usually adapt when ridden by a sensible owner and taken out with a reliable horse as escort to start with.

Don't expect the seller to let you take the horse out on your own—would you trust your horse to a total stranger? Instead, you should be given the chance to go out with a sensible rider, ideally on a route which gives you the chance to meet traffic and have a short canter on suitable going.

Don't just plod along behind your escort horse all the time. By all means start off behind, but if the horse seems calm and well-behaved, ask if you can take the lead to make sure that he is happy to be in front too. If your route includes a short stretch where you can canter, let your escort go first and see how your horse reacts—he'll probably be eager to keep up with his friend, but hopefully he won't pull your arms out trying to race him. Ideally, you should then swap positions and ask your horse to go in front for a while, but this may not always be possible to do so.

By the time you get back to base, you should know whether or not you would feel happy with this horse. If you've brought an advisor with you, have a chat about your observations and findings; if you've come alone and want a second opinion, arrange to come back as soon as possible. The seller may tell you that other people are coming to look at the horse, which may or may not be true, in which case the sensible course is to stand firm and insist on a second visit. There is, of course, the risk that in the meantime someone else will try and buy the horse, but that disappointment would be far less than buying in haste and then finding out you've made a wrong decision.

Some people may suggest that you ask to have the horse on trial for a week or two to see how you get on with him. Few owners would be prepared to do this, because of the risks involved—so while there's nothing to prevent you asking, don't be surprised or suspicious if the seller declines.

Final Checks

Once you've found the perfect partner, it's sensible to ask a specialist equine vet to check him over and make sure he's sound and suitable for the job you want him to do. You may not like the thought of spending even more money, but it's well worth it. A vet will check everything from the horse's heart, eyes, and lungs to whether he remains sound when his joints are flexed and he is asked to trot away afterwards.

RANGE ROVER
3 DZ

A new partnership

When you start out on a relationship with your new horse, don't be in too much of a hurry. The first few weeks should be a process of getting to know each other and you also need to give your horse time to settle in to his new surroundings. Some horses settle quickly, while others may take weeks before they are totally at home.

Your first problem may be how you are going to transport him. You may find that the seller is able to deliver him for you, but if not, you may have to rely on an experienced friend or a professional transporter. Don't attempt to do it yourself unless you are already experienced in towing a trailer with a horse on board and are aware of the necessary techniques.

Just because you can drive a car, it doesn't mean you can tow a trailer, even though you may be legally entitled to do so. It takes practice to be able to give a horse a smooth, safe ride and transporting a load which can move about isn't as easy as transporting one which remains static.

Try and make the changeover from your horse's home to another as smooth as possible by following his previous regime as closely as possible—even if you intend to change it. For instance, if he is getting hard feed, buy the same brand to start with and if he is used to hay, don't suddenly switch him to haylage, and vice versa. If possible, ask the seller to let you have a couple of bales of forage so that you can gradually mix one supply with the other.

Make sure that a stable is ready for him so that when he arrives there will be plenty of bedding laid down, clean water, and forage. This will give you a safe environment in which to remove his traveling gear and let him have a look around his new environment. If all the other horses are likely to be turned out in the field when he arrives, try and arrange for at least one quiet horse or pony to be stabled nearby, where he can see him, as this will give him reassurance that he isn't alone.

If he arrives in daylight, you may want to turn him out straight away so that he can have a roll and graze. Ideally, a new arrival should go out in a small field with a quiet companion and if necessary, moved to a larger area when he has settled. If there is the slightest risk that he could have a health problem or been in contact with a horse who has—which may be the case if he comes from a stable where there is a constantly changing equine population—the stable owner may want him to be kept so that he does not come into contact with others for a short time.

Such precautions are only worthwhile if anyone who handles the horse does not come into contact with others at the stable. However, the horse in quarantine should still be able to see other horses or he will become distressed.

While it may seem kind to give him a week or two to settle in before you start riding him, this can be counter productive, as he may have rather too much energy for comfort when you get on him. If he seems reasonably settled the next day, try and turn him out first thing to let off any steam and plan to ride or lunge him later in the day, if time and daylight hours allow.

You're bound to be excited at the idea of riding your new horse for the first time; you may also be a little nervous. This is quite understandable, especially if he is your first horse, so you might want to arrange for your teacher to be there to give you confidence. Again, this will be a "getting to know you" session rather than an intensive schooling one.

When you ride out for the first time, arrange for a sensible friend with a sensible horse to go with you. Try and find a quiet circular route that will keep you out for about an hour, allowing for the fact that you will be walking most of the time with short periods of trot when the going permits.

Hopefully, you'll soon start to build a partnership, but don't be disheartened if you get the occasional blip. Your horse might not understand what you're asking him to do, or he might be wary of new things in a strange environment. Think positive, remember that your horse relies on you for direction, encouragement, and support—and have fun!

Enjoying horses

If you can't own, borrow, or share a horse, there are still plenty of ways to enjoy them. Riding clubs and branches of the Pony Club welcome enthusiasts who don't have their own mounts, and you'll find plenty of ways to become involved in their activities, such as helping at competitions and rallies.

Take every chance to watch other riders at local and national competitions. Don't just watch them in the ring: study how they prepare their horses beforehand in the collecting ring (warm-up area). Many riders give lecture demonstrations and these are a great way of getting an insight into riding and training.

There are holidays on horseback for all ages, abilities, and budgets. Wherever you go and whatever you do, whether it's taking an intensive course at an equestrian center or trail riding in an exotic location, spending time with horses and trying something new—even if it's only for a week or two—will increase your confidence and ability and, hopefully, give you great pleasure.

Horses have a special magic, whether they be top-class competition animals that have become household names in their own right or family ponies. There's no such thing as an ordinary horse, so make the most of every encounter.

153

PART 7

APPENDICES

Glossary

AHS: African horse sickness.
aids: signals given by the rider to aid communication, such as leg pressure, rein pressure, and voice commands.
American barn: internal stable set within a large building.
Azoturia: old colloquial name for equine rhabdomyolysis syndrome.
back at the knee: a conformation fault in which the outline of a horse's leg between the knee and the fetlock is concave.
bib clip: the hair on a horse's chest, the underside of his neck, and throat is clipped off.
bit cheeks: these come in several designs, pass through the ends of a bit mouthpiece, and are the attachment point for the reins.
bit rings: free-running rings that pass through the ends of a bit mouthpiece and are the attachment point for the reins. An alternative to bit cheeks.
blanket clip: the hair is clipped off to leave just a blanket of hair over the horse's back.
blaze: a wide white stripe down the center of the face.
bone: the term given to a measurement taken around the widest part of the cannon bone, just below the knee.
bradoon: a thin snaffle that forms part of a double bridle.
breastplate: or breastcollar, or breastgirth; an arrangement of leather straps attached to the saddle and girth to prevent the saddle from sliding backward.
browband: the bridle strap that crosses the forehead.
brushing: term used when a horse touches or knocks one leg against its partner on the same side.
cannon bone: long bone between the knee or hock and the fetlock.
cantle: back of a saddle seat.
cast: term describing a horse that's lain down near a wall and become jammed against it when he tries to get up.
cavesson: a type of noseband.
chaser clip: the hair is clipped off following a diagonal line from the ears to the end of the belly, leaving hair on the back and hindquarters.
cheekpieces: the bridle straps that run down the sides of the horse's face.
cinch: name used for girth securing a Western saddle.
clenches: the turned down ends of the nails that hold horseshoes in place.
cob: a chunky, short-legged, deep-bodied horse type.
Coldblood: a horse belonging to the heavy/draft breeds.
colic: any sort of abdominal pain.
colt: an uncastrated male horse up to three years old.
conformation: the shape and build of a horse; a horse or pony with good conformation is one that meets standards recognized as being helpful for soundness and athleticism.
coronet: the band of tissue immediately above a horse's hoof.
cow hocks: term used to describe hocks that turn inward; a conformation fault.
cow-kick: forward kick using a hind leg.
crib biting: or cribbing; stereotypical behavior in which a horse seizes a ledge, door, or fence post in his teeth.
crossover: another name for the Grakle noseband.
croup: the rump or hindquarters of a horse.
curb: a lever-action bit.
curb chain: a chain attached principally to a curb bit, which passes under the chin.
dishing: term used to describe a horse turning one or both front limbs out to the side as he moves.
dock: the solid part at the base of the tail.
double bridle: one which employs two bits to give refined communication between an experienced rider and a well-schooled horse.
drop: a type of noseband that passes across the lower part of the face and fastens below the bit.
EDT: equine dental technician.
eel stripe: a dorsal stripe running down the back.
EIPH: exercise-induced pulmonary hemorrhage.
entire: or stallion; an uncastrated male horse over three years old.
equine rhabdomyolysis syndrome: a disorder of the hindquarters, similar to muscle cramp.
ergots: callouses underneath the fetlocks.
ermine spots: spots of black coloring in a horse's socks or stockings.
ERS: see *equine rhabdomyolysis syndrome.*
ewe-necked: a horse that looks as if his neck has been set on upside down.
feathers: long, silky hair on the legs of many heavy horses and some ponies.
fenders: type of stirrups used on Western saddle.
fetlock: leg joint between the cannon bone and pastern equivalent to the human ankle.
figure-of-eight: another name for the Grakle noseband.
filly: female horse up to the age of three years.
Flash: a type of noseband that fastens above and below the bit.
Flehmen response: see *vomeronasal organ.*
fly ring: the loose ring in the center of a curb chain.
forearm: the upper front leg.
forging: term used to describe a horse clipping the back of a front foot with the toe of a hind foot as he moves.
French link: a variety of double-jointed snaffle bit.
frog: the V-shaped cleft in a horse's foot.
full clip: or hunter clip; all the hair is clipped off except for that on the legs.

gag snaffle: snaffle bit with slots in the cheeks to take running bridle cheekpieces.
Galvayne's groove: the vertical line on the corner incisors of a horse more than nine to ten years old.
gelding: castrated male horse.
girth: the depth of a horse's body; also the strap used to hold a saddle in place.
going close in front/behind: terms used respectively to describe a horse's forelegs or hind legs coming close together as he moves.
GP saddle: general purpose saddle suitable for hacking, schooling, and jumping.
Grakle: a type of noseband.
grass sickness: a disease thought to be caused by a soil-borne bacterium called *Clostridium botulinum*.
gridwork: fences set at prescribed distances to improve a horse and rider's accuracy and confidence.
gullet: the deep groove on the underside of a saddle to allow clearance of the spinous processes.
hack: a type of horse used for hacking.
hackamore: variety of bitless bridle.
hacking: trail riding.
halter: see *headcollar*.
hand(s): measurement used in describing the height of a horse from ground to the highest point of the withers; one hand = 4 inches (10cm). Fractions are traditionally rendered as fourths of a hand, traditionally meaning inches, although measurements are now often given in centimetres: thus a horse described as 12.2hh (or hands high) is 12 hands and 2 inches tall (128cm).
headcollar: or halter, or headstall; an arrangement of straps or ropes around a horse's head by which it can be led or tethered.
headpiece: the bridle strap that goes behind the horse's ears.
headstall: see *headcollar*.
heavy horse: often used as a generic term to describe draught breeds bred to work the land.
hh: hands high.

hock: the largest hind leg joint.
hogged mane: one where all the mane and forelock hair has been "hogged," or clipped off.
horse walker: a mechanical circular pen with sections to take two or more horses, which turns at slow speed and is used to exercise horses at walk with no rider or handler.
Hotblood: term once widely used to describe Thoroughbred and Arabian horses.
HPA: hoof pastern axis; the angle of the dorsal hoof wall and the pastern.
Hunter: a horse that's used for any form of hunting.
in hand: led, not ridden, as in trotted in hand to assess movement or soundness.
Jacobson's organ: see *vomeronasal organ.*
kimblewick: a variety of bit.
laminitis: a painful foot condition.
livery yard: a professionally run facility where horses can be kept and trained.
loose box: alternative name for a stable building in which a horse has freedom to move around.
lozenge snaffle: a variety of double-jointed bit.
lunge cavesson: a headcollar with a reinforced noseband into which rings are set as attachment points for a lunge rein.
lungeing: training or exercising a horse by moving him in a circle round his handler, who controls him by means of a lunge rein attached to a lunge cavesson or to the bit.
lunge whip: a long whip used to indicate speed and direction to a horse being lunged.
mare: female horse over three years old.
martingale: a strap designed to prevent a horse raising his head above the angle of control.
mud fever: a bacterial skin infection.
mullen: a bit with an unjointed, slightly arched mouthpiece.
noseband: the bridle strap that encircles the horse's nose.
over at the knee: a conformation fault where the outline of a horse's leg between the knee and the fetlock is convex.
overreaching: term used to describe the way some horses strike the back of a front foot with the toe of a hind foot when they move.
overshot: term used to describe a horse whose upper jaw is longer than his lower jaw; also referred to as a "parrot mouth."
parrot mouth: term used to describe a horse whose upper jaw is longer than his lower jaw; also referred to as "overshot."
pastern: the part of the leg between the hoof and the fetlock.
pelham: a variety of bit.
plaiting: term used to describe a horse putting one front foot in front of the other as he moves.
points: collective term for the different parts of a horse's body.
pommel: front arch of a saddle.
port: an arch in the middle of a bit mouthpiece.

quartering: grooming a horse a quarter at a time.
rack: a high, fast Saddlebred gait.
rain scald: a bacterial skin infection.
RAO: recurrent airway obstruction.
riding horse: a horse that is in effect a cross between a hack and a hunter.
ringworm: a fungal infection of the skin.
roach back: term used to describe the outline of a horse's back when it's convex.
roached mane: American term for hogged mane.
saddle patch: patch of hair left on a clipped horse in the area where the saddle sits.
scawbrig: variety of bitless bridle.
sclera: a white ring all the way round a horse's eye.
set-fast: old colloquial name for equine rhabdomyolysis syndrome.
sewing machine trot: a short, up and down stride.
sickle hock: term used to describe a conformation fault in which a hind leg is in front of a perpendicular line dropped from the hock to the ground.
sidepull: variety of bitless bridle.
slow gait: a high, four-beat Saddlebred gait.
smegma: waxy substance produced in the sheath and penile area by geldings and stallions.
snaffle: the simplest form of bit, but with many variations.
snip: a white marking on the muzzle that may extend into the nostril.
socks: white leg coloring anywhere up to the knee or hock.
splint: bony growth on the inside of the cannon bone.
stallion: or entire; an uncastrated male horse over three years old.
star: an irregularly shaped white patch on the center of the forehead.
stifle: the hind leg joint above the hock, near to the flank.
stirrup bars: part of the saddle from which the stirrup leathers and irons are suspended.
stockings: white leg coloring that extends up to and over the knee or hock.
strangles: a highly contagious respiratory disease, most often seen in young horses.
stripe: a narrow white stripe down the center of the face.
surcingle: a strap or straps which secure a blanket.
sweet itch: an allergy to the saliva of the biting midge *Culicoides*.
tack: generic term for all saddlery equipment used in connection with a riding horse; usually used to refer to the saddle and bridle.
tack up: put the saddle and bridle on a horse.
throatlatch: or throatlash; the bridle strap that runs beneath the horse's throat from ear to ear.
trace clip: the hair from the underside of the neck and throat and the lower part of the body is clipped off.
training aids: equipment used to try to influence a horse's head carriage and posture.
tree: the frame on which most saddles are constructed.

tushes: vestigial canine teeth of male (and occasionally female) horses.
tying-up: old colloquial name for equine rhabdomyolysis syndrome.
undershot: term used to describe a horse whose upper jaw is shorter than his lower jaw.
Vomeronasal organ: or Jacobson's organ; a sac at the top of the nasal passages that is used to investigate unusual or stimulating scents by means of a behavior known as the Flehmen response, in which the horse raises his nose and curls his top lip. This partly closes off his nostrils, enabling the organ to analyse the scent more fully.
walleye: a blue eye.
Warmblood: a horse whose breeding combines a mixture of coldblood and hotblood influences.
weaving: stereotypical behavior in which a horse stands and repeatedly sways his head from side to side.
whorl: a pattern of hairs radiating out from a central point.
wind sucking: stereotypical behavior in which a horse repeatedly gulps down air.
withers: the bony rise at the base of the neck.
WNV: West Nile virus.
wolf teeth: small, shallow-rooted vestigial molars found in many horses.

Useful contacts

American Association of Riding Schools
8375 E. Coldwater Road
Davison, MI 48423-8966
www.ucanride.com

American Farriers Association
4059 Iron Works Parkway
Lexington, KY 40511
www.theamericanfarriers.com

American Association of Equine Practitioners
4075 Iron Works Parkway
Lexington, KY 40511
www.aaep.org

Association of British Riding Schools
Queen's Chambers
38–40 Queen Street
Penzance, Cornwall TR18 4BH, UK
www.abrs-info.org

British Equine Veterinary Association
Wakefield House
46 High Street, Sawston,
Cambridgeshire CB2 4BG, UK
www.beva.org.uk

British Horse Society
Stoneleigh Deer Park
Kenilworth, Warwickshire CV8 2XZ, UK
www.bhs.org.uk

Registry Resources

Akhal-Teke
The Akhal-Teke Society of America, Inc.
P.O. Box 207
Sanford, NC 27331
www.akhaltekesocietyofamerica.com

American Azteca
American Azteca Horse International Association
P.O. Box 1577
Rapid City, SD 57709
www.americanazteca.com

American Bashkir Curly
American Bashkir Curly Registry
857 Beaver RoadWalton, KY 41094
www.abcregistry.org

American Cream Draft
American Cream Draft Horse Association
193 Crossover Road
Bennington, VT 05201
www.acdha.org

American Creme Horse
American White and American Creme Horse Registry
90000 Edwards Road
Naper, NE 68755
http://awachr.com/home

American Miniature Horse
American Miniature Horse Association
5601 S. Interstate 35 W.
Alvarado, TX 76009
www.amha.org

American Miniature Horse Registry
81 B Queenwood Road
Morton, IL 61550
www.shetlandminiature.com

American Paint Horse
American Paint Horse Association
P.O. Box 961023
Fort Worth, TX 76161
www.apha.com

American Quarter Horse
American Quarter Horse Association
P.O. Box 200
Amarillo, TX 79168
www.aqha.com

American Saddlebred
American Saddlebred Horse Association
4083 Iron Works Parkway
Lexington, KY 40511
www.asha.net

American White Horse Horse
American White and American Creme Horse Registry
90000 Edwards Road
Naper, NE 68755
http://awachr.com/home

Andalusian
International Andalusian and Lusitano Horse Association
101 Carnoustie North, Box No. 200
Birmingham, AL 35242
www.ialha.org

Appaloosa
Appaloosa Horse Club
2720 West Pullman Road
Moscow, ID 83843
www.appaloosa.com

International Colored Appaloosa Association, Inc.
P.O. Box 99, Shipshewana, IN 46565
www.icaainc.com

AraAppaloosa
AraAppaloosa Foundation Breeders' International
Route 8, Box 317
Fairmont, WV 26554

Arabian
Arabian Horse Association
10805 E. Bethany Drive
Aurora, CO 80014
www.arabianhorses.org

Belgian
The Belgian Draft Horse Corporation of America
125 Southland Drive, P.O. Box 335
Wabash, IN 46992
www.belgiancorp.com

Belgian Warmblood
Belgian Warmblood Breeding Association
North American District, 1979 CR 103
Georgetown, TX 78626
www.belgianwarmblood.com

Brabant
American Brabant Association
2331A Oak Drive
Ijamsville, MD 21754-8641
www.theamericanbrabantassociation.com

Brindle
International Buckskin Horse Association, Inc.
P.O. Box 268
Shelby, IN 46377
www.ibha.net

Brindle and Striped Equine International
11819 Pushka
Needville, TX 77461
www.geocities.com/sbatteate/brindlehos

Buckskin
International Buckskin Horse Association, Inc.
P.O. Box 268
Shelby, IN 46377
www.ibha.net

Canadian
Société des Éleveurs de Chevaux Canadiens
Canadian Horse Breeders Association
59 rue Monfette Suite 108
Victoriaville, Quebec G6P 1J8
Canada
www.lechevalcanadien.ca

Chincoteague Pony
National Chincoteague Pony Association
2595 Jensen Road
Bellingham, WA 98226
www.pony-chincoteague.com

Cleveland Bay
Cleveland Bay Horse Society of North America
P.O. Box 483
Goshen, NH 03752
www.clevelandbay.org

Clydesdale
Clydesdale Breeders of the USA
17346 Kelley Road
Pecatonica, IL 61063
www.clydesusa.com

Colonial Spanish
Horse of the Americas, Inc.
202 Forest Trail Road
Marshall, TX 75670
www.horseoftheamericas.com

Connemara
The American Connemara Pony Society
P.O. Box 100
Middlebrook, VA 24459
www.acps.org

Dales Pony
Dales Pony Association of North America (Canada)
P.O. Box 733
Walkerton, Ontario, N0G 2 V0, Canada
www.dalesponyassoc.com

Dales Pony Society of America
32 Welsh Road
Lebanon, NJ 08833
www.dalesponies.com

Dartmoor Pony
American Dartmoor Pony Association
203 Kendall Oaks Drive
Boerne, TX 78006

Dartmoor Pony Registry of America
295 Upper Ridgeview Road
Columbus, NC 28722
www.dartmoorpony.com

Drum Horse
Gypsy Cob and Drum Horse Association, Inc.
1812 E. 100 N.
Danville, IN 46122
www.gcdha.com

Dutch Warmblood
The Dutch Warmblood Studbook in North America
P.O. Box, Sutherlin, OR 97479
www.kwpn-na.org

Exmoor Pony
Exmoor Pony Association International
P.O. Box 1517
Litchfield, CT 06759
www.exmoorpony.com

Exmoor Pony Enthusiasts
P.O. Box 155, Ripley, Ontario, N0G 2R0
Canada
http://exmoorenthusiasts.fortunecity.com

Canadian Livestock Records Corporation
2417 Holly Lane
Ottawa, Ontario, K1V 0M7, Canada
www.clrc.ca

Falabella Blend Miniature Horse and Falabella Miniature Horse
Falabella Blend Registry
33222 N. Fairfield Road
Round Lake, IL 60073-9636
www.falabellafmha.com

Fell Pony
Fell Pony Society
Ion House, Great Asby,
Appleby, Cumbria CA16 6HD
UK
www.fellponysociety.org

Fell Pony Society and Conservancy of the Americas
775 Flippin Road, Lowgap, NC 27024
www.fellpony.org

Florida Cracker Horse
Florida Cracker Horse Association, Inc.
2992 Lake Bradford Road
S. Tallahassee, FL 32310
www.floridacrackerhorses.com

Foundation Quarter Horse
Foundation Quarter Horse Registry
P.O. Box 230
Sterling, CO 80751
www.fqhr.net

Friesian
The Friesian Horse Society, Inc.
17670 Pioneer Trail
Plattsmouth, NE 68048
www.friesianhorsesociety.com

Friesian Heritage Horse
Friesian Heritage Horse and Sporthorse International
133 E. De La Guerra, #159
Santa Barbara, CA 93101
www.friesianheritage.com

Bibliography

Bowers, Nathan, and Katie Bowers Reiff. *4-H Guide to Training Horses*. Voyageur Press, 2009.

Budiansky, Stephen. *The Nature of Horses*. Weidenfeld and Nicholson, 1997.

Devereux, Sue; edited by Karen Coumbe. *The Veterinary Care of the Horse*. J. A. Allen, 2006.

Diggle, Martin. *The Novice Rider's Companion*. Kenilworth Press, 2009.

Henderson, Carolyn. *Getting Horses Fit*. J. A. Allen, 2006.

Henderson, Carolyn. *Horse Tack Bible*. David and Charles, 2008.

Henderson, Carolyn. *Bring Out the Best In Your Horse*. J. A. Allen, 2009.

Henderson, John. *Glovebox Guide to Transporting Horses*. J. A. Allen, 2004.

Hyde, Dayton O. *All the Wild Horses*. Voyageur Press, 2009.

Johnson, Daniel, and Samantha Johnson. *How To Raise Horses: Everything You Need To Know*. Voyageur Press, 2007.

Johnson, Samantha, and Daniel Johnson. *The Field Guide to Horses*. Voyageur Press, 2009.

Lynghaug, Fran. *The Official Horse Breeds Standards Guide: The Complete Guide to the Standards of All North American Equine Breed Associations*. Voyageur Press, 2009.

MacLeod, Clare. *The Truth About Feeding Your Horse*. J. A. Allen, 2007.

Index

Acknowledgments

The author and photographer would like to thank the following owners and producers for help with photography: Anne James and Tim Northcott, owner and producer respectively of Morgan stallion Landside Music Lord; Maria Lucas, owner of the Fjord, Savanna Gretchin; Mandy Morgan, owner of the Haflingers Oxnead Archie, Oxnead Marisa and Schieferstein Amberleon; Jo Taylor and Hazel Taylor, owner of partbred Appalosas Centyfield Humdinger and Centyfield Hero. Special thanks to show producers Lynn Russell and Kate Jerram and to Georgia and Lauren Maynard and Robbie the pony.

First published in 2010 by Voyageur Press, an imprint of
MBI Publishing Company, 400 First Avenue North, Suite 300, Minneapolis, MN 55401 USA

First published in Great Britain in 2010 by
Haynes Publishing Group, Sparkford, nr. Yeovil,
Somerset BA22 7JJ, England.
This edition published in 2010 by Voyageur Press, MBI Publishing Company.

ISBN-13: 978-0-7603-3940-4

Voyageur Press Editor: Fran Lynghaug

Printed in China